The
Mind
of
an
Ape

By Ann J. Premack

Why Chimps Can Read   (1975)

By David Premack

Intelligence in Ape and Man   (1976)

W. W. Norton

&

Company

New York

London

David
Premack
Ann
James
Premack

The Mind of an Ape

Copyright © 1983 by Ann J. Premack and David Premack

Published simultaneously in Canada by George J. McLeod Limited, Toronto.
Printed in the United States of America.

The text of this book is composed in Laurel, with display type set in Stymie.
Composition by New England Typographic Service, Inc.
Manufacturing by The Murray Printing Company
Book design by Elissa Ichiyasu

First Edition

Library of Congress Cataloging in Publication Data
Premack, David.
The mind of an ape.
Includes index.
1. Apes—Psychology.   2. Primates—Psychology.
3. Mammals—Psychology.   I. Premack, Ann J.   II. Title.
QL737.P96P736 1983   599.88'0459   82–18796

ISBN 0-393-01581-5

W. W. Norton & Company, Inc., 500 Fifth Avenue, New York, N. Y. 10110
W. W. Norton & Company, Ltd., 37 Great Russell Street, London WC1B 3NU

1 2 3 4 5 6 7 8 9 0

We wish to thank the taxpayers of the United States for supporting this research through the National Science Foundation.

Contents

The
Mind
of
an
Ape

Introduction ■

The
Hairy
Golem

■ No one has ever believed that a dog raised in a human home would one day turn into a child. The dog is too different from us to provoke such an assumption. Early domesticators of the chimpanzee, however, often assumed that if they brought an ape into their home, hugged it, bottle-fed it, and swaddled it in diapers, it would one day become "almost human." Indeed, it was thought that, under the right circumstances, the ape might even acquire language.... An enriched environment and the proper instruction might work near-miracles. Had not John B. Watson himself promised: "Give me a dozen healthy infants ... and I'll guarantee ... to train [them] to become ... doctor, lawyer, artist, merchant, chief, and, yes, even beggar-man thief...."

The earliest attempt to transform a chimpanzee failed, it was thought, because language was erroneously equated with speech. Perhaps the chimpanzee could not make human language sounds because of the construction of its vocal tract. And this, many believed, was only a superficial problem. If one were to find another form of language, using signs or symbols, might not such a language yet be acquired by the chimpanzee? The manipulative hands of the ape could succeed where its lips and tongue had failed.

The sense of similarity which we feel intuitively toward the chimpanzee is confirmed by the sciences of genetics and evolution. Geneticists tell us there is a ninety-nine percent overlap between the chromosomes of human and chimpanzee; a far greater overlap than between horse and zebra, or even chimpanzee and monkey. In addition, physical anthropologists, who reconstruct fossils, have found evidence that the ancestors of monkeys separated from those of humans at least twenty-five million years ago, while those of chimpanzees and humans separated a mere five million years ago.

Our initial sense of likeness is based on superficial features, features which describe primates: eyes in the front of our faces, hands rather than claws, and the ability to stand erect for some part of our day. Simple as are these similarities, they set monkeys, apes and humans apart from all other creatures. But there is another similarity, psychological rather than anatomical, which sets the human and chimpanzee apart even from the monkey.

Should we gaze at a monkey, it would reveal no inclination to exchange glances; it would furtively avert its eyes, or enlarge them in a hostile stare. Monkeys are threatened by the gaze of others. Apes and humans, in contrast, enjoy being the focus of attention. Humans regard being looked at as an honor, as proof of personal achievement. Far from averting its eyes, the chimpanzee returns our gaze avidly. While looking at us, the chimpanzee appears to be asking the very questions about us which we ask about it as we gaze into its eyes.

The notion that a similarity exists among members of the primate group will never be granted by some people, no matter how tidy the scientific evidence, or how careful the direct observation. Most of us, however, find that genetics and anthropology only verify the obvious. What is essential, we believe, is a better understanding of the "obvious." What do species have in common? More important, in what ways are they different? For, though we can clearly see the similarities, we are not blind to the differences. If we can clearly state how we differ from our closest relatives, we begin to make a contribution to the question of who we are. That's our objective.

Speech, of course, sets humans apart from chimpanzees, and emphasizes what is both natural and universal to the human species. Humans acquire language in every kind of environmental setting. Even the most primitive of peoples, in the most isolated areas, have language. But language is not the only competence natural to humans. We also analyze the world into actions which have causes and effects, attribute states of mind to others, and represent the world in our heads—a capacity which permits us to walk around with our knowledge even when the world is not visible. These competences are natural, untaught; as natural to humans as the ability to speak.

Our own attempt to teach language to an ape was not based on the belief that a chimpanzee could be turned into a child. We were interested in the human mind. In order to understand the mind of the human, it was essential, we thought, to compare it with other minds. But were there other minds that were comparable?

Some thought there was essentially only one kind of mind in the universe, shared by all creatures; that the minds of others differ from the human's in the way that, say, mere houses differ from a castle—the same floor plan but fewer rooms, lower ceilings, and so on. Some argued that the human mind was quite like that of others except for the addition of a room called "language." Some insisted that the addition of language completely transformed the house, so that the difference was no longer that between a house and a castle, but between a house and an airplane. Thus, no proper comparisons could ever be made.

How could we choose among these basic views about mind without having determined the nature of language itself? But even today, after twenty years of brilliant progress in linguistics by scholars around the world, our knowledge of language is insufficient. In 1954, the year our research began, ignorance concerning language was almost painful. We had no idea what a human might be like without language. The only way to find out was to deprive a child of its language environment—an unthinkable experiment. It was possible, however, to do the opposite—to teach language to an ape. At least to try.

We began, then, to teach a simple, written language, one we invented, to a creature that in the ordinary course of its natural life does not acquire a language. For the subsequent fifteen years we also examined many capacities other than language that might differentiate the mind of one creature from that of another. Can the chimpanzee manage abstract reasoning, for instance, analogies, such as, chimpanzee is to human as house is to castle? Could it be trained to read maps? To do simple arithmetic? Learn to lie? To deceive others who are intent on deception? In the course of answering these questions, we began to uncover previously hidden capacities in the mind of the ape.

While this book emphasizes the language progress made by our oldest chimpanzee, Sarah, twenty years old at this writing, it also follows the development of eight other chimpanzees: Elizabeth, Peony, and Walnut, who were all students of the invented language, and Jessie, Sadie, Bert, and Luvie, who were not. The eighth, Gussie, accompanied Sarah from Africa and was reared with her for several years. Though systematically exposed to the language as Sarah had been, Gussie failed to learn a single word.

All of our animals were born in the wilds of Africa and entered the laboratory when they were quite young. We acquired, first, Sarah and Gussie; later, Jessie, Sadie, Bert, and Luvie as mere infants. All were bottle-fed and diapered for several years. We did not raise the chimpanzees in our home. In the early days, we did, on occasion, bring the very young Sarah and Gussie to our house for a visit. Gussie, always anxious, clung to her security blanket and needed to be carried; but Sarah played with the children, "watched" television, and played tag with the dog. The children were delighted by Sarah. Never before had their normal house rules been so quickly and thoroughly violated. We think it unwise to raise the chimpanzee at home as either pet or surrogate child. Not only is the chimpanzee a poor pet, but when raised as a potential child, its every act will be anthropomorphized, undermining the scientific value of the animal.

There is a tradition in psychological laboratories, especially in medical ones, to house animals in virtual solitude. Since we are interested in intelligence, in the mind, we prefer to maintain animals in an environment that is complex and stimulating. Only under these conditions can the diverse activities of a species remain normal. In a static environment animals fall into a kind of torpor. Even their appetites decline.

Nor did we starve or abuse our animals. Some experimenters in behavioral psychology keep animals in solitude, then counteract the apathy that results by starving the animal, and then requiring it to "work" assiduously to regain some part of the deprived food. Sometimes the animal is given doses of electric shock if it responds incorrectly—or too slowly. This is hardly an astute procedure for exploring a mind. A starved animal is not necessarily a more thoughtful one, and an animal that has been given electric shock may be gripped more by fear than by the need to solve problems.

Our young chimpanzees were cared for quite differently and had, in fact, rather a nice life. In 1964 when Gussie and Sarah came to Missouri, the two lived comfortably in an apartment above David's rat research laboratory on campus. While the rats downstairs pressed levers for sugared water, the chimpanzees upstairs were being diapered and bottle-fed by a surrogate mother. They slept in our children's outgrown baby cribs (a roof of sorts had been added in order to prevent them from roaming at night). When we moved, the cribs came along with the chimpanzees to the University of California at Santa Barbara, where the animals were welcomed at the airport with a warm bottle and a change of diapers.

The front lawn of the psychology building at Santa Barbara soon became the new playground for Sarah and Gussie. Both chimpanzees ran freely and climbed the trees, many of which, like the chimpanzees themselves, were natives of Africa. Occasionally, hordes of students from an anthropology class at the University came to have a look at the chimpanzees' teeth in full, live display. Except for the brief occasions when they were held immobile for such dental demonstrations, the chimpanzees were unrestrained.

Probably their finest hours were spent on a bluff overlooking the Pacific Ocean, where they ran wildly on the hard-packed sand, often startling dogs into flight with their sudden bursts of speed. The gallop of the chimpanzee is astounding. This animal, which sometimes surprises us by walking upright (Sarah once seized a dog's leash and took the panicked dog for a leisurely stroll, the essence of a matron on her way to church), can also gallop on all fours like a demon. The same powerful, brachiating arms that swing the animal from tree to tree become another set of legs, thrusting the animal forward at a fiercesome pace. Sometimes on these jaunts, Sarah would discover a tattered jacket or pair of torn pants, some odd article of clothing left behind by a student. The clothing may have evoked a nostalgia for the shirts and pants she once wore as an infant, for she never failed to seize these remnants and struggle into them. After fighting her way into the jacket, she would sail along the beach, the long trailing garment hiding her legs—a dark apparition moving with startling speed.

In the wild, chimpanzees eat mainly fruit and leaves. In the laboratory, they are maintained on commercial biscuits (a balanced diet by themselves), fresh fruits, and vegetables. These staples are enlivened by tidbits—nuts, dried fruits, and yogurt—doled out when lessons end.

When young, Sarah was always tested in her home cage, space akin to a two-room apartment, in which both trainer and Sarah sat at the same table (though more often Sarah sat on the table). The intimacy of this arrangement helped the lesson over its occasional hard parts, allowing the trainer to coax and cajole her student to try "just a few more times." The trainer would pat Sarah's hand or chuck her beneath the chin, and Sarah accepted this as lavish praise for her correct answers. Sometimes Sarah reciprocated, chucking her trainer.

At sexual maturity, when Sarah gained the strength of an adult chimpanzee, it became advisable for the trainers to remain outside

FIGURE 1: *The laboratory and compound were designed by Esherick, Holmsey, Dodge and Davis, a San Francisco firm. When the lab was moved to Pennsylvania, the original design was retained, except for minor adaptations to accommodate the colder climate.*

the cage. This diluted the intimacy, and, for a while, lessons were fractious and unreliable. But as time went by, Sarah adapted to the new arrangement, and lessons resumed normally. Although trainers sat outside the cage, safe from Sarah's dangerous arms, which she could swing forcefully when in a rage, Sarah and the trainers still managed to engage in mutual grooming. Sarah invited grooming by turning her back and glancing over her shoulder at the trainer; the trainer in turn extended her leg or foot, human body parts that Sarah liked to groom. The leg seems prized for its hair, the foot for its toes.

Food and housing are less important in the daily life of the chimpanzee than are the social relations between the animal and its trainers. In the wild, the chimpanzee is not weaned until four or five years of age and it may remain with its mother until puberty. When removed from the wild and brought to the laboratory, the chimpanzee transfers its maternal ties to the human trainer. David

FIGURE 2: *Recently transplanted from California, Peony was quite willing to don sweatshirt and socks before taking a stroll in the much cooler Eastern countryside.*

discovered how deeply chimpanzees feel about humans on his first job as a research associate at the Yerkes Laboratories of Primate Biology in Orange Park, Florida. Each day, he carried the chimpanzees he was testing from their home cage to the test area, where all the experimental materials were stored.

One young, female chimpanzee used to go willingly to be tested but was far from willing to return to her home cage. When David opened her test door to fetch her home, she would bare her teeth and rock back and forth, her fury mounting. On one occasion, she escaped from her test space and ran wildly around the larger test room. David pursued her hotly. Brandishing a broom, he finally drove her into a corner. She lingered there momentarily and then sprang at him—"to sink her teeth into me," David anticipated. But, no. She clung to him like a truant child, trembling in his arms.

FIGURE 3: *Five-year-old Peony behaved well with her trainer. The leash was used mostly to mollify strangers.* PHOTOGRAPH BY BRUCE STROMBERG.

Frightened by the "raging mother" she had herself antagonized to fury, the chimpanzee, now contrite, tried to soothe her own fears and her "mother's" anger by capitulating.

Our present laboratory, located some miles from the University of Pennsylvania, and designed originally to be built on the California coast, was modified by its San Francisco architects to accommodate a pond rather than an ocean view. It nestles amid lush Pennsylvania-Dutch farmlands, facing a tree-lined pond that is dramatically open to view from all offices and meeting rooms. The gentle slope of land in back of the building is surrounded by what, from a distance, appears to be a football stadium but is in fact the outdoor compound. When in the compound, while reclining on the rugged beams of their towering swings, the animals can survey a breathtaking panorama of oak, pine, and the verdant Amish farmlands. Here the animals play and forage on the plants and insects; Bert, the male, bold enough to live in a haven of females, is also

FIGURE 4: *In good weather the animals spend the entire day playing, fighting, and foraging in the compound. Many of the tests, such as the ones on natural reasoning (see pages 110–13), are also conducted in the compound.*

facile enough to capture and kill the field mice that populate the compound. Laboratory animals need not be housed in barren cages—certainly ours are not.

First to arrive here from Santa Barbara were Sarah and Peony, escorted by one of their California trainers aboard a cargo plane. While Peony was soon delighted with our move, Sarah paced about irritably. Peony, who was younger and more tractable, dressed in her colored socks and boxer shorts every afternoon to be taken for a walk along the country roads. Her trainer held her in check with a long leash (although no leash could have contained Peony had she wished to be free, for Peony was huge). Frequently, the two promenaders would encounter an Amish cart drawn by a spirited, delicate horse. The driver, bearded, black-hatted, "plain-clothed," would rein abruptly. Intense mutual staring followed. It was never clear who was more fascinated: Peony, the Amish driver, or the horse.

During our first year in Pennsylvania, we lived temporarily in a trailer on the laboratory grounds, and Peony would, at the end of her walk, knock on Ann's door to request cookies and pickles. Sometimes she came in for a more social visit—and perhaps tea with her cookies. But Sarah had no such prerogatives after she reached puberty. Less tractable, sometimes treacherous, and much older (she was then twelve to Peony's five), she was confined indoors to her large duplex cage with its picture-windowed view of the pond and countryside.

Perhaps Sarah was at a loss to understand how a seemingly intelligent group of people could have left the golden California climate for heavy eastern humidity. In a fit of pique, she bit Ann's finger. Ann recalls the incident mainly because it occurred shortly before Thanksgiving, a holiday usually celebrated with a stove laden with pots and pans. David prepared that year's banquet. It was his ape, after all, who had done the damage. Sarah never showed the slightest remorse for her act.

Almost everyone who has worked closely with chimpanzees has been bitten. Bites are reminders that chimpanzees are not domestic animals, no matter how tame they appear. They are wild animals, able to injure even those they like most—a difficult lesson for those who pride themselves on having gained the "affection" of the chimpanzee.

Today, at almost twenty years of age, Sarah cannot be taken outdoors for walks or into special rooms for testing. Everything is brought to her. Her cage is connected to an outdoor run that overlooks the pond. Her trainers bring a TV set and phonograph into her room every day; they prepare her tests and offer her several classes a day, not unlike tutors indulging a special student with private lessons five days a week. Sarah was always constitutionally short-tempered and demanding; with increasing age, she has become even more of a crank. Now, she must "ring" for her tests. Five times each day, the buttons on a panel board in Sarah's room light up briefly. If Sarah pushes a button when the board is lit, the trainer enters with a test appropriate to the choice of button. This procedure ensures an interested Sarah. Fortunately, her periods of balkiness have been reduced. There is nothing more disturbing to graduate students' morale than having their lessons received with a huge yawn, which Sarah does not bother to conceal.

Our four juveniles—Bert, Luvie, Sadie, and Jessie—live very differently. They have been raised as a group since infancy, and even now, though seven years of age, they can be taken to and from the compound where they play and forage in good weather. They, too, are tested several times a day. Called in from the compound by name, they come immediately when summoned by the trainer. "Bert!" "Luvie!" brings the animal called. They are remarkably cooperative about coming in for their classes. Trainer and chimpanzee, hand in hand, walk companionably down the hallway to the "classroom." Thirty minutes later a restive "student" is returned outdoors to recess, while "Sadie" or "Jessie" respond to a new call.

Although these animals were never raised in a human home, one might say they have been given the advantages of a private boarding school. They receive the daily lessons and have the devoted caretakers that are lavished usually only on a child. Do these advantages transform the chimpanzee? If so, how?

There is a tradition in Jewish mysticism concerning the creation of a golem, a creature constructed from inert matter—one who can even understand speech.

Jakob Grimm describes the legend as it existed in 1808: "After saying certain prayers and observing certain fast days, the Polish Jews make the figure of a man from clay or mud, and when they pronounce the miraculous Shemhamphoras [the name of God] over him, he must come to life. He cannot speak, but he understands fairly well what is said or commanded. They call him golem and use him as a servant to do all sorts of housework. But he must never leave the house. . . ."

The methods that the mystical books, the Kabbalah, provide for the creation of the golem reflect the perceived power of language. No act had greater power than speech, in particular the speech of God. And quite as God "spoke over and into" the clay in creating Adam, so could a human speak God's name over a clay figure to create a golem. Not surprisingly, when humans create a golem, they are usually more successful in producing a merely animate creature. To create a golem that can comprehend speech is a more difficult task.

We now know that someone who comprehends speech must know language, even if he or she cannot produce it. But in that time, the early nineteenth century, only those who spoke were considered to have the power of language. The true power. The ability to create with words as God created the world. By simple declaration.

Early Judaism demeaned *the image* on the one hand and exalted *the power of language* on the other. God was equated with the "word," and threats to God were equated with idols and graven images. This tension suggests that in the formation of an early monotheism, the word had not yet fully won its battle over the image for the possession of the human mind. The modern mind credits both speech and image, but language may not always have been in the ascendant position. At one time, the dominant way of transmitting information may have been through images.

While word and picture once competed in the battle for the human mind, the word now stands ascendant. In this book we will once again encounter minds in which a battle between word and image rages. But here, the outcome is different. This time, in these other minds—in these hairy golem—the word is not destined to win.

Chapter One

■

A Language Designed for an Ape

■ Children in every part of the world acquire language in much the same order, moving from words to complex sentences as a chrysalis turns into a butterfly. The transition from chrysalis to butterfly is simple, however, compared to the child's passage from silence to speech. If we place a cocoon in a bottle, the chrysalis will not fail to become a butterfly—that is its genetic destiny. However, if we isolate a child, "bottle" it, essentially, the child will not emerge with spoken sentences. The metamorphosis of language depends on both genetics and experience—though we still do not fully understand how the two interact. We know that mothers do not drill their children in speech. We also know that unless the child and mother communicate, the child will not develop speech at all.

The chimpanzee is not genetically destined to speak any more than humans are destined to fly. Apes do not develop speech by simple exposure to a speaking "mother"; but perhaps, with drill,

FIGURE 5: *In 1967 Sarah was about five and a half years old. She is shown here at her magnetized language board, handing a word back to Mary Morgan, her favorite trainer. In the beginning of her language training, Sarah usually had two sessions daily, lasting about one-half hour each.*

they can develop, if not speech, then a specially designed ape language. Could we not devise a language so simple, a training program so forceful, that the combination would play the role of genetics? We would drill the ape in language the way a teacher drills children in skills they do not acquire naturally—the three R s, for example.

If we could do this, we would have at least two advantages. First, in developing a simple language and comparing it to our complex human one, we would learn something about language. Second, if the ape itself could be taught, we could question it, as we question a child, and learn what the ape understands of its world.

The language system we invented for the chimpanzees was written rather than spoken. The elementary unit of the invented language was the "word," a piece of colored plastic. Each word had a particular color and shape—a small, blue triangle was the word for apple; a small, pink square, the word for banana. This novel language could be seen but not heard and touched but never pronounced. The words did not vanish as do spoken words; each plastic word was backed with metal and adhered to a magnetized slate. Sentences did not appear and disappear in time but were "written" piece by piece, in a vertical sequence that remained on the language board for however long was needed for careful scrutiny (see illustration 6 on page 18).

Our first language student, Sarah, was five years old when she first enrolled in school. Her lessons were simple in those sunny days. A seated instructor would place some bit of food on the table, and Sarah, seated across from him or her, would take the food and eat it. The exchange soon became a routine activity. When the instructor tried, on occasion, to request some of the food that Sarah had taken by either extending a cupped hand or extruding her lips in supplication (mimicking the requests for food that chimpanzees use with one another), nothing happened. Sarah might seem to hesitate momentarily, as though puzzled, but "giving back" was not a transaction in which she willingly participated. She preferred always to be the recipient.

A vital first step in teaching language is to establish a social exchange between the chimpanzee and the language instructor. In their natural environment, chimpanzees engage in grooming, giving and taking, eating, copulating, greeting—the many social exchanges that they carry out in their daily life. Taking, however, is not a frequent social exchange, for the simple reason that wild chimpanzees do very little giving. Indeed, what we might call giving is more often "tolerated taking," where one animal begs insistently from the other until the latter gives in. But in the laboratory we can increase the frequency of this social exchange: if the trainer serves as giver, the chimpanzee cooperates by taking. Teaching language with a routine such as giving should be effective, we reasoned, since there appear to be no concepts in this routine that are beyond the animal's understanding. Why should the chimpanzee not recognize the elements involved: the person giving; the action of giving; food;

FIGURE 6: *Elizabeth, one of the three animals exposed to language training, is shown here "writing" a simple request. Blocked from view, the top word is "Debby" (the trainer's name); the next two words are "give" and "Elizabeth." To complete the construction, Elizabeth must add the name of an item she wants, for example, "apple" or "banana." Notice that the pendant around Elizabeth's neck, her name, corresponds to the name on the board.*

her taking; herself? Our objective was to teach her a word for each separate part of giving.

When the giving routine was well established, the actual mapping with the language system began. The trainer placed a word, the piece of colored plastic, on the table along with the corresponding piece of fruit, a banana. The name for banana, however, was more easily within Sarah's reach than the banana itself. The chimpanzee was encouraged by example and coaxing to place the word on the language board. The action of placing the appropriate plastic word on the magnetic slate was Sarah's language equivalent of our uttering the word "banana."

We started with fruits, offering different fruits during the feeding exchanges with a corresponding change in words for each fruit.

When the fruit was a banana, the plastic chip was of one kind; when an apple, the name was a different piece of plastic; and when an orange, the name was changed once again. During every feeding exchange, Sarah's task was the same: to place the piece of plastic, proferred with the fruit, on the magnetized slate. She would then be given the fruit by the trainer.

We followed that experiment with a similar one using the names of donors. To better identify them, donors wore their plastic names like pendants on a necklace. Sarah had her own to wear. When Mary was the donor, Sarah had to place Mary's name on the board; when Randy was the donor, it was Randy's name that had to be placed on the magnetized board. Now the chimpanzee had to correctly name both the donor and the fruit in order to be given the fruit. In other words, the animal could no longer request the fruit by writing "apple" but had to write "Mary apple" ("Randy banana"), placing both the name of the donor and the fruit on the board. In addition, Sarah was required to observe a particular order in writing two or more words. "Mary apple" was accepted, but "apple Mary" was not. It was now clear that the chimpanzee could recognize both the person giving and the fruit. We next taught Sarah words for the other parts of the giving routine: the action of giving and the recipient. In this way, Sarah reached the target sentence "Mary give apple Sarah."

This account of Sarah's instruction in her language is highly idealized. We had numerous problems along the way. Some arose from Sarah's limitations, others from our own errors. Sarah had considerable difficulty learning her first words; later, having learned the names, she sometimes named a fruit she desired rather than the one that was presented. Difficulties also arose in teaching Sarah names for such ideas as "if-then," "some," "all," and so on. To find simple examples of these not-so-simple ideas was not always easy—and sometimes we unwittingly misled Sarah.

An observer, seeing the trainer hold a plastic word-form alongside some object, demonstrating to Sarah in this casual way that the plastic form was the "word" for the object, often could not believe that it took Sarah (as well as Elizabeth and Peony) hundreds of trials in order to form the first associations between the plastic words and objects.

Some of our apes have not learned a single word—Gussie, Sarah's original companion, was one of these—and other chimpanzees may take a thousand trials before learning their first word. Why is it so difficult for an ape, a higher primate, to make an association between a blue plastic triangle and an apple? It seems incongruous, especially since we know that a pigeon would probably manage the association after only a few hundred trials.

In testing an animal's ability to associate a word with an object, we place an apple on a table. Below it, we place two plastic words: a blue triangle and a pink square. The animal need only place the appropriate word, the triangle, on its magnetic board. Simple, it seems. Not for the chimpanzee. During the thousand trials when she continues to perform randomly—putting the pink square on the board as often as the blue triangle—is she learning nothing?

Hardly. The animal is learning the function that these two kinds of objects serve in the test environment. We know, for instance, that the plastic words can buy the animals fruits that they can eat. But does the animal know this? Indeed, if, at a time when it is still failing to make the correct associations, we scatter many plastic words and pieces of fruit on the ape's work table, the ape will spontaneously hand the trainer a plastic word for each piece of fruit. While these are not the names we have assigned for the fruit, the animal does not hand over one piece of fruit for another one or, for example, try to eat a word. It knows that words get you fruit to eat. The ape has also learned which four words and which four fruits are part of its problems—for, if given a new word or a new fruit on tests, the chimpanzee will always reject both. In addition, since in the beginning, everything for which we tried to teach a name was a fruit of one kind or another, the animals came to know the general kind of thing for which words could be used. For instance, although the animal definitely preferred chocolate to fruit in its home cage, when offered the same choice in the classroom, the animal always requests the fruit with a word; it has learned that in school, words get you fruit, not chocolate.

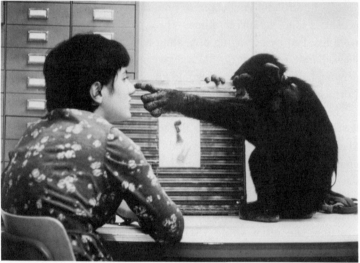

FIGURE 7: *Peony carries out the instructions that her trainer has written on the board. The sentences shown here read, "Peony banana insert" and "Peony nose touch."*

Thus, while the animal takes hundreds of trials to learn that a pink square is associated with a banana, and a blue triangle with an apple, something the pigeon would learn in fewer trials, we find that this period is not wasted. The chimpanzee is learning about classes of items (fruit and words) and their functions (for eating, for requesting) before finally learning specific associations. The pigeon, of course, can learn the associations of pink square with banana and blue triangle with apple rather speedily. But it learns little else.

After the first fruit names were learned, other fruit names and the names of donors were easily mapped. But the names of recipients other than "Sarah" and actions other than "giving" met with complications, which while not impossibly knotty, were unexpected. One in particular was amusing. Sarah was reluctant to write "Mary apple Gussie" on the board, since this meant Mary would now give the apple to Sarah's cagemate. Sarah sturdily resisted permitting anyone other than herself to be a recipient. Actions such as "wash," "cut," "insert" also encountered some resistence, for Sarah clearly preferred some actions to others. These strictly practical problems were handled by arranging appealing contingencies. For instance, when Sarah would write "Mary give apple Gussie," generously denying herself an apple, we would give Sarah a bit of chocolate. If "altruism" is given a handsome payoff, it can become quite reliable.

That Sarah had some conception of word order can be seen from her very different reactions to the instructions "Sarah give apple Mary" and "Mary give apple Sarah." When an instruction called upon Sarah to be the donor, she might obey at first, but ultimately she would scream and knock the words off the board. On one occasion, she took her name and stamped it vigorously all over the pieces of apple (as if to say, "Mine! Mine!"). With or without instruction, she often wrote "Mary give apple Sarah" but never "Sarah give apple Mary."

A rule of sequence that requires the giver to be named before the receiver is needed because any individual can give to any other. If only females were donors and males receivers, for instance, we need not say "Mary give apple John." We could just as well say "John give apple Mary." But where no such semantic rules exist, we need rules of sequence that dictate which word comes first.

Prepositions such as *on, under,* and *beside* involve physical relations among which, we do not doubt, the chimpanzee can discriminate. The preposition *on* is like "giving" in that any object can be *on* any other; even as any individual can give to any other. At least in the special world in which Sarah was language trained, this was the case. She was given a variety of colored cards, any one of which might be *on* any other—rather than objects such as dishes and fruit, an asymmetrical case in which fruits are placed in dishes but not the reverse.

Having taught Sarah plastic words for the colors red, green, blue, and yellow in earlier sessions, we used them for training the symmetrical "on" relation. We placed a red card on Sarah's table, wrote "red on green" on her board, then induced her to place a red card on the green one. We followed this instruction with "green on red," and she placed a green card on the red. Subsequently she was given both cards and required to behave correctly according to her changing instructions by placing the red card on the green one or the green on the red. When she was proficient, she received all four colored cards at the same time. She responded correctly not only to "red on green" and "green on red," but also to "green on yellow," "yellow on red," and so on, demonstrating that she understood the word "on." Sarah was next given the opportunity to describe the position of the cards with her words. She did very nicely. When given the color names, she wrote spontaneously what she had been trained to do: "red on green" when the red card was on the green one, "yellow on red" when the yellow card was on the red one, and so forth.

Same or Different:
Is That a Question?

Sarah was able to learn names for the various individuals who both gave and received, for objects that exchanged ownership, and for

FIGURE 8: *Represented here are early "same" and "different" constructions. Consisting of both objects and words, they were called "hybrids" and were written on the table. Later constructions, of words only, were written on the board. The words for same and different while of the same shape, varied in other features, most notably in color—orange and red, respectively.*

the action of giving. Was this because the animal's natural request, which is commonly found in the wild, can be so easily translated into an imperative type of sentence? Chimpanzees extend their palms to others when seeking to obtain food. Were Mary eating an apple, an uneducated Sarah would request food with an upturned palm; the educated Sarah, then, could easily translate the gesture into the imperative construction "Mary give apple Sarah." In fact, a semi-educated Sarah, while in the early phase of word training, made an imperative-type request. One morning, food pellets still lay on Sarah's work table since language lessons had not yet begun. Her trainers were drinking coffee. Sarah picked up one of the food pellets and extended it to one of the trainers in an act that was not a supplication, resembling instead a demand. The trainer immediately understood, accepted the pellet and gave Sarah a sip of coffee; Sarah offered another pellet, received another sip, and the exchange continued until the coffee was gone.

Since the success of the imperative seems to be based on its natural counterpart (the requests chimpanzees make in the wild), were we destined to fail in teaching Sarah the concept of *question?* There would seem to be no natural counterpart for it in the wild. In attempting to teach the question to Sarah, we needed to find a basic concept (or set of them) around which to build the interrogative type of sentence. The concept of same/different proved remarkably useful in this regard.

Sarah was given identical objects placed slightly apart on her

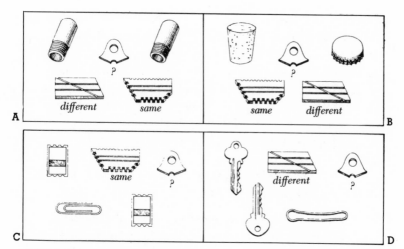

FIGURE 9 A-D: *Here are Sarah's first questions having to do with the concept of same and different. A and B ask, "Are these two objects the same or different?" Sarah had to choose the correct answer. In C and D the question posed is, "What object is the same as (or different from) the object that is shown?" Sarah had to choose the correct object.*

worktable and received the orange plastic word "same" to be set between them; she was then given her red plastic word "different" to place between two unlike objects (perhaps a cup and a spoon). Had Sarah formed the desired associations? To find out, we presented two alike objects (two cups) and gave her the words for same and different simultaneously. She was required to choose the correct one. On other trials, we gave her unlike objects (a cup and a spoon) and both words, again requiring her to choose between them. In addition, she received a variety of objects that were previously unknown to her to identify as same or different. Her performance was so close to perfect on all the tests that, in principle, one could picture her going about her cage, picking up an infinite variety of pairs of objects, and labeling them "same" or "different" with ease.

Sarah's identity exercise as described above is, in a sense, already a question. The verbal equivalent of the above task would be something like, "What is the relation between these two objects? Are they the same, or different"? An English-speaking person would have little problem understanding the question or grasping the nature of the answers. The chimpanzee's format does not contain an explicit interrogative marker, but the schema is implicitly a question. We made it explicit by introducing a piece of plastic to stand for a question mark. Now we placed a question mark between the two cups (and the cup and the spoon), making the questions explicit.

Sarah was presented with a variety of interrogatives that are best translated into English as follows:

What is the relation between one clothespin and another?
What is the relation between a clothespin and a pair of scissors?

Sarah replied to such questions by removing her interrogative marker and substituting her words for same and different in reply.

When it was clear Sarah knew the words same and different, we changed the form of the question. We removed one of the objects, replaced it with a question mark, and asked, "Clothespin same?" and "Clothespin different?" That is, we asked her what object was the same as a clothespin and from what object it differed. Of course, the alternatives offered were not words but objects of several kinds. Once again, Sarah replied to these questions by removing the interrogative marker and substituting the correct object in reply.

There is yet another form of the question we taught Sarah, to which the reply was either yes or no. In this kind of question, the interrogative marker did not stand for the absent words for same and different (or a missing object). Instead, it stood for yes or no, as in the following:"? clothespin same clothespin," "? apple same banana," "? clothespin different clothespin," "? apple different banana." The correct replies are yes, no, no, yes, respectively. These new words, yes and no, were not used spontaneously. They, too, were taught Sarah in the same arduous way as all her other vocabulary.

"No" had been taught earlier in the form of a double instruction: "Sarah take jam cracker, No Sarah take peanut butter bread."

FIGURE 10: *Sarah was taught the meaning of "no" with pairs of constructions. The sample shown here translates as "Sarah jam bread take" and "No Sarah honey cracker take." When Sarah complied by reaching for the bread and jam, she was allowed to take it. When she reached for the cracker and honey, the trainer restrained her. Sarah soon learned to refrain from items in negated constructions.*

When Sarah reached for the slice of bread, she was prevented from taking it. She rather quickly formed an association between no and her tendency to reach for the negated food. She soon learned to take the permitted food but suppressed even reaching for the forbidden, or "negated," food.

Although the question was introduced in the context of same and different, it did not remain associated with these two words only. Sarah understood the broader implication of the question mark to be the missing information that she was to supply. Missing information later came to include such concepts as color of, shape of, size of. She could then be asked "? color of apple," "small ? raisin," "round shape of ?"—to which she chose the alternatives red, size of, and ball, respectively.

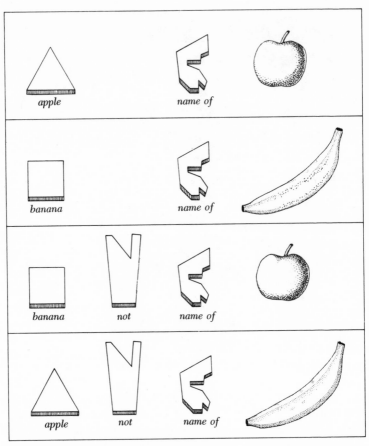

FIGURE 11: *The top two constructions are: "apple name of a real apple" and "banana name of a real banana." The bottom two are instructions that "banana not name of a real apple" and "apple not name of a real banana." Since we use actual objects (banana and apple) to make the process of naming explicit, these are also hybrid constructions.*

Imperative sentences resemble the natural requests made by apes. And Sarah learned the imperative construction by making requests of others and by carrying out requests others made of her. Interrogative sentences, in contrast, do not seem to resemble any natural form of ape behavior, yet Sarah appears to have learned their meaning. Is this so? Though she understood the question, she did not herself ask any questions—unlike the child who asks interminable questions, such as What that? Who making noise? When Daddy come home? Me go Granny's house? Where puppy? Sarah never delayed the departure of her trainer after her lessons by asking where the trainer was going, when she was returning, or anything else.

She did delay her trainer's departure once by stealing the words. Seating herself on the floor in the middle of her cage, she crouched over her words and wrote out meticulously the same kinds of questions her trainer had previously asked her. Her replies were generally correct. We allowed this theft of the words to become a routine. Sarah could then play with the words used in every particular lesson, imitating the questions given her and a moment later answering them.

The child wants to know the name of a person, wonders if a dog bites, inquires where her Daddy went. For she recognizes a deficiency in her own knowledge and is as deeply impelled to correct this deficiency as she is to watch a cartoon or to chew bubble gum! Since Sarah managed very well to pose the question constructions to herself (not to mention the answers to them), her failure to ask questions was not based on an inadequacy at the level of the form of the question. The ape's failure is due to its inability to recognize deficiencies in its own knowledge. Language training can supply the form of the question, but it cannot teach a creature to examine the state of its knowledge or to find deficiencies that impel the desire for information.

Metalinguistics

In her early training, Sarah learned to distinguish between plastic words and objects. Only later did she learn to associate a specific plastic word with an object. While she was taught many relations between words and objects, she was never taught, specifically, that her colored plastic chips "named" objects. So we introduced the

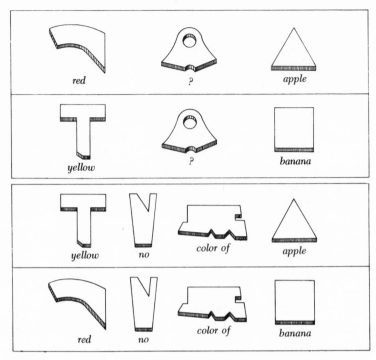

FIGURE 12: *Sarah originally learned the question mark (interrogative particle) in constructions that asked if objects were the same or different (like those in Figure 9). She could also understand when asked to identify the relation between an object and its color: "What is the relation between red and apple?" and "What is the relation between yellow and banana?" She could even be asked more complex questions to which her replies were: "Yellow no color of apple" and "Red no color of banana."*

plastic word "name of," which made it explicit that her language consisted of words for things.

We wrote "Blue triangle ? actual apple" and gave her the new plastic word for "name of." She removed the question mark, substituting "name of" in its place, writing "blue triangle name of actual apple." Next, we wrote "Pink square ? actual apple" and offered "not-name of" to her. "No" (taught her in an earlier lesson) was glued to the word for name of; Sarah once again removed the question mark and added the newly offered word to write "Pink square not-name of apple." Next, to make sure Sarah had formed the desired associations, she was given both "Blue triangle ? actual apple," with "name of" and "not-name of" as alternatives, and "Pink square ? actual apple," with the same alternatives. This is, of course, the procedure that we used in teaching the words "same" and "different." We continued to follow the procedure but now asked her questions involving words and objects that had not been used in training, such as "Raisin ? actual raisin," "raisin ? actual apricot." The answers were "name of" and "not-name of," respectively. We even went further and asked her, concerning an apple, "? apple name of real apple" and "? apple name of real orange"; Sarah replied yes to the first and no to the second question.

Sarah's success inclined us to even greater inquisitiveness about her ability. We wanted to see if she understood the naming procedure well enough to be able to use the new word "name of" to acquire the name of a *new* object. We gave her the instruction "fig name of actual fig" and "crackerjack name of actual crackerjack"; both were foods she liked but had not yet learned to name. Then we wrote "fig not-name of actual crackerjack" and "crackerjack not-name of actual fig." To see if she understood these instructions, we asked her such questions as "crackerjack ? actual fig"; she replied "not-name of" correctly. She prompted us now, by her performance, to give her the acid test. Placing a ripe fig on the table, we gave Sarah several words—the new words for fig and crackerjack, along with some of her old words: give, Sarah, Mary, orange, and banana. Sarah correctly wrote, "Mary give fig Sarah," in the presence of an actual fig, and "Mary give crackerjack Sarah," when the crackerjack was present.

Internal Representation:
A Blue Triangle Becomes a Word

As we look at Sarah's dictionary, a large box of colored pieces of plastic with English definitions, we are led to wonder when a pink square (or a blue triangle) ceases to be simply a perceptual form

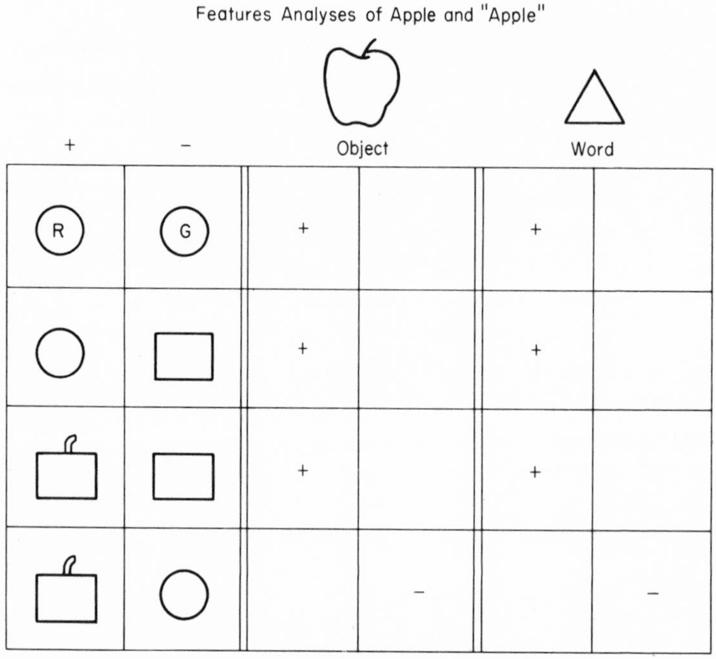

Features Analyses of Apple and "Apple"

FIGURE 13: *On the left of the figure are the pairs of characteristics from which Sarah could choose in order to describe an actual apple; she chose the same features to describe her word for apple as she did to describe a real apple, even though the word for apple was a blue triangle.*

and enters the realm of word. Partly, we conclude, it is after Sarah uses the pieces of plastic appropriately, for example, requests apple with the blue triangle, and when asked "what is the name of an apple?" replies with the blue triangle. This reflects the standard kind of definition of a word, and though we cannot improve this view of word use, we would like to add another dimension in terms of our language system: the piece of plastic becomes a word when the properties ascribed to it are not those of the plastic but of the object it signifies. In looking at the word, one sees not it, but the object for which it stands. Because Sarah's words were plastic forms, it was easy to test what Sarah saw when she looked at her words.

First, we showed her an apple and asked her to describe it by giving her the following alternatives: a red card versus a green one, a square card versus a round one, a square card with a stem on top versus one without the stem, and finally a square card with a stem versus a round one without the stem. After obtaining Sarah's features analysis of the apple, which turned out to be red (not green), round (not square), square with stem (not square without stem), and about seven choices of round without stem to every three choices of square with stem, we asked her to describe her word for apple. She was given the same alternatives in the presence of the blue plastic triangle. She ascribed to the word all those features she had previously chosen for the apple itself, indicating that her blue plastic triangle was red, round, square with a stem; she even repeated her seven-to-ten split in favor of round over square-with-a-stem. For Sarah, then, a blue plastic triangle represented a red, round object with a stem, quite as the word for Mary does not stand for a capital *M* followed by three lowercase letters but represents a blonde woman who was Sarah's favorite trainer.

While many variations remain to be explored, the basic procedure for teaching words to a naïve organism is extremely simple. An exchange is established between the animal and the trainer, and the trainer decides what parts of the exchange can be distinguished by both trainer and student. In the *giving* exchange, for instance, both parties can distinguish the donor, the recipient, the action, and the object transferred. When a new member of the donor class is introduced, a corresponding change is made in the word. The same procedure holds for the action, recipient, and object classes. In this manner, the animal acquires words, as well as combinations of words, that permit it to understand and build ever larger constructions: for example, "apple," "Mary banana," "Randy cut fig," "Debby give peach Sarah." When Sarah writes, for instance, "Randy give grape Sarah," we have acid proof that Sarah indeed does understand all parts of the giving exchange. Now, the educated Sarah can describe, in words, the very exchange in which the uneducated Sarah was but a mere participant.

This kind of language is clearly very different from ours. Every word we write is based on every word we speak. For Sarah, there was no spoken or heard language on which to base her writing. She did not have the free use of her words in the sense that children can talk freely; they have no need of pencils to make their sentences. Sarah could only exchange information when her words were physically available. Chimpanzees, we now know, are not initiators of language; they will, however, engage in dialogue once drawn into an exchange. It was this ability, through the system of language that Sarah learned, that gave us an invaluable tool for investigating a far more important issue, the mind of the ape.

Chapter Two

■

From Simple Judgments to Analogies

■ At an early age, children divide the world into things that do and those that do not look, feel, smell, and taste alike. For example, when playing with a set of blocks, they will place the blue ones to one side and the red ones to the other side. Later, they may redivide the same blocks into categories of wood and plastic, again placing each group to different sides. They do this natually, spontaneously, without instruction or training. Though not as strongly disposed to categorize the world, chimpanzees also sort objects according to color, size, and so on.

When Sarah was about eighteen months of age, she received an unexpected caller. A friend of Sarah's caretaker dropped by for a short visit. The woman happened to be wearing a distinctive plaid skirt, woven of thick wool. Sarah, usually hesitant with strangers, stepped boldly toward her visitor, stroked and handled the woman's skirt. Suddenly, she walked over to her crib and began to stroke and handle her thick wool blanket, which was also plaid.

Not yet trained in language, Sarah could not say the word for same. She celebrated her discovery of an equivalence in the only way open to her, by feeling the texture of the two like items in the same interval of time.

Placing like items in the same location or touching them in the same interval of time—spontaneous categorization—is not an idle game. Species that do spontaneous categorization tend to be capable, as well, of learning a kind of language. The act of placing together things that look alike may seem no more than judging whether they are "same" and "different," but such is not the case. The first act (placing alike items together), depends entirely on resemblance—on how things look, feel, smell, etc.,—whereas the second depends on considerations deeper than that of superficial similarity. We will explain the difference more fully in the next few paragraphs.

"Same" or "different," as a judgment, is so basic to the way we perceive our world that we have difficulty describing events without it. We use the judgment unthinkingly in describing our own behavior and that of birds and bees. For instance, in observing a pigeon select grain, we assume the bird judges each new particle to be the same as one it has already eaten (in which case it eats the particle), the same as one it has already rejected (in which case it avoids that particle), or different from either (in which case it explores or tries out the particle). Yet, it has not been demonstrated that the pigeon is, in fact, making such judgments. The judgment of same/different is more complicated than meets the eye. Rather demanding tests must be designed in order to establish that another creature judges in terms of same or different. Examples of such tests are described below.

The Judgment of
Same/Different

The match-to-sample procedures used with Sarah are only a first step in determining whether a chimpanzee can make a same/different judgment.

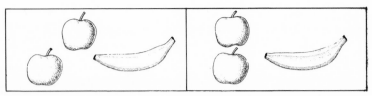

FIGURE 14: *On the left is the triangular arrangement used in match-to-sample tests. The sample is at the apex of the triangle and the two alternatives are at the base. On the right is the correct solution; the matching alternative is placed below the sample. Putting like objects together is a primitive form of responding. It is a common activity in young children and is found also in young chimpanzees.*

The procedures we taught Sarah involve simple objects used in pairs, such as cups, spoons, clothespins, rubber bands, postage stamps, pipes, cotton balls, keys, paper clips, and blue beads. Sarah's first problem consisted of a cup, as the sample, and a cup and spoon as alternatives. The problem was presented in the form of a triangle, the cup (the sample) displayed at the apex of the triangle with the alternatives at either base. Initially, Sarah was guided through the correct matching movements. Her large hand was placed on the cup resting at the base, then both her hand and the cup were moved gradually until they reached the sample. The guiding procedure was repeated after reversing the positions of the alternatives, so the cup alternative was as often on the right (the spoon on the left) as on the left (with spoon on the right) of the triangle's base. This balancing of the alternatives informs the ape it must pay attention to more than a particular position when choosing an alternative.

Sarah received several trials using a cup as the sample, with cup and spoon as alternatives, and several trials using a spoon as the sample, with cup and spoon as alternatives. Her errors were few. Since her performance was so promising, she was shifted to a set of transfer tests using the group of simple objects listed above, that is, the clothespins, rubber bands, keys, paper clips, and so forth, all of which were novel to the test format.

Yet Sarah was perfectly able to match a whole new set of objects that were alike, even though she had never been tested on them before. These results made clear that she had learned more than simply to put cups together or to place spoons together. She had adopted a rule, essentially informing her that in this test format, she should place the look-alikes together.

The ability to match items in a general way is a laboratory version of what a chimpanzee does in its natural environment—sort things spontaneously.

Reacting to the similarity between objects is in fact simple, and many species can manage it. Sarah, other apes, pigeons, and other non-primates can all perform well on generalized match-to-sample testing. Once a pigeon has been trained to match two colors, it can then match not only other colors, but also two alike shapes. But if the bird transfers from matching colors to matching shapes, how can we speak of "reacting to similarity" since colors bear no resemblance to shapes? The answer to this puzzle lies in our knowledge of the nature of learning.

In the basic experiment in learning, even a simple creature such as a pigeon, learns associations not only on one level but on several. For instance when we train a bird to peck the red alternative in the presence of a red sample, it might learn: (1) approach red, (2) match red to red, (3) match colors, (4) match all items that resemble one another (regardless of their dimension or modality). Which of these are viable alternatives for the bird?

All of them are. Although final tests remain, our educated guess is that the bird learns on all levels. Notice that each level becomes more abstract than the one before. The least abstract, the first level, concerns the color actually before the bird's eye. Since the color is also before the experimenter's eye, the bird's ability to learn to approach red is not surprising. The third level, that of matching colors, still involves the external world; we too can see that the bird is matching colors. But by the time we reach the fourth level, what the bird has learned no longer refers to anything that is visible. We must assume, however, that the bird has learned on this level, too, for only if it has, are we able to explain the bird's ability to transfer to the matching of shapes after learning to match colors.

Performing generalized match-to-sample tests is simply a higher order of responding to physical similarity. While obviously the color red does not look like a square shape, the relation between one square and another is no less a case of physical similarity than is the relation between one red color and another: if we learn to match things that look alike, we will place a square with another just as readily as we place a red card with another card (which is red).

Both the pigeon and the ape perform generalized match-to-sample tests by responding to physical similarity—they differ, however, greatly in the relative efficiency. The ape learns matching almost immediately, then proceeds to demonstrate perfect transfer; whereas the pigeon learns slowly, and its transfer is weak. We can explain this phenomenon perhaps, by returning to the fact that learning occurs on many levels. An individual's resources must be divided over all the levels, and the division need not be equal. With the pigeon, the division favors less abstract levels, so the bird learns mainly to approach color (or whatever else is used in training). Hence matching is learned slowly, and its transfer is weak. The ape, in contrast, devotes its resources to the more abstract levels, and thus learns matching quickly, and its transfer is stronger. The ape not only divides its psychological pie differently, it probably has a larger pie to divide.

A true same/different judgment goes far beyond responding to mere physical resemblance. One may place one apple with another because of the resemblance, likewise one elephant with another because of their resemblance. But what about the two elephants with the two apples? To make this judgment, we can no longer compare one element with another. Moreover, we cannot now rely on physical similarity or resemblance. Two apples do not in any way look like two elephants. Nevertheless, they clearly belong together; for both are cases of sameness.

In match-to-sample tests, we can use relations—e.g., pairs of elements—as both the sample and the alternatives. For example, we may use AA as the sample, BB and CD as alternatives; as well as AB as sample, CD and EE as alternatives. Now the individual must match BB to AA (for they are both cases of sameness), as well as match CD to AB (for they are both cases of difference).

There is little chance the pigeon will pass a test requiring it to match relations; for even three- to four-year-old children find the problem difficult and require special training. The ape is in a similar predicament. It cannot pass the test wthout special training. That training is in language. When so trained, the ape passes the test, but not before.

As a part of Sarah's language training, we taught her to use the words "same" and "different."

Once again, we presented Sarah with her original cup problem, but now within a new context. The cups were placed in a horizontal line, and a piece of plastic meaning "same" was placed between the cups. When a spoon was substituted for one of the cups, the word for different was placed between these unlike objects. After Sarah ran through several of the above errorless trials (so called because the procedure arranges for a consistently correct outcome), she was moved to a choice procedure, which gave her the opportunity to demonstrate what she had learned. She was offered the luxury of making errors.

Sarah was given two cups spaced apart and both "same" and "different" as choices to place between them; following that were a cup and a spoon with the same word choices. What had Sarah learned from the errorless trials? Quite a bit, since she made few errors with "same." She made more errors on the "different" trials, but the number was not excessive. Had they been more abundant, she would have been returned to the round of errorless trials as a review, then restored once again to her tests.

Sarah's success at this level meant she could progress to our "transfer" tests, where she would be examined on the same test format but with a host of novel objects. She had, now, to decide whether a host of new and old objects—mixed and changed freely—were the same or different. Sarah made few errors even on these tests, indicating she had learned more than "put same between cups," "put same between spoons," "put different between cup and spoon," and so on. She seemed to have learned rules rather like "put same between look-alikes" and "put different when things don't look alike," although the latter rule seemed uncertain and more subject to interference.

Peony, who was far less adept at learning the problems that Sarah managed so handily, engaged in what is closer to the labeling of like and unlike objects. Because we had so much difficulty teaching same/different to Peony and had restricted all her training to the triangle format, we wondered whether she had the general idea of same/different or could perform only in that one format. To find out, we placed three objects and two words in a paper bag and simply emptied the contents onto Peony's worktable, thus completely destroying the fixed format. Given this challenge, Peony demonstrated the nontrivial and, indeed, remarkable character of ape intelligence.

She resolved the chaos into her own order, represented by several innovative arrangements. While a variety of arrangements could have been constructed with the three objects (a piece of clay and two spoons or vice versa) and the words for same and different, Peony demonstrated that, in spite of her seemingly tenuous grasp of same/different (she needed numerous trials compared with Sarah), her comprehension was more than superficial.

First, she placed the two spoons together and set the word for different on the portion of clay. In this arrangement, "same" was left out, as though it was redundant for this analysis. In another trial, she placed the two spoons together, adding the word for same on top of them and placing the word for different beside the portion of clay. In yet another solution, she wrote out, in linear fashion, "A piece of clay—same—a piece of clay—different—a spoon." After many paper-bag trials, Peony finally settled on her preferred format for the illustration of same/different. She placed the two alike objects on top of one another, with "same" balanced precariously above the pile, and set the word for different beside, or on top of, the unlike object.

Proportion: A Unifying Aspect
of Unlike Objects

The ape is quite able to make same/different judgments about objects that look alike. But what if objects do not look alike; what if they have only some aspect in common? For instance, could the chimpanzee match such aspects as small, spotted, half, quarter when the objects to be compared bore no resemblance to one another except in that one aspect being considered? Would the chimpanzee match a small ball to a small triangle rather than to a medium-sized ball? A red cup to a red ribbon rather than to a white cup? A half orange with a half-filled bottle rather than a whole orange?

Sarah had been tested on colors, sizes, and shapes, as well as many other aspects of objects in the language-training procedures, but she had never been tested for knowledge of proportion. So we tested, not only Sarah, but also the four juvenile chimpanzees (the young animals who had never been taught Sarah's language system) on their ability to judge proportions in a match-to-sample test format (the same test triangle used in the same/different tests). We used both solid and liquid proportions in this series of tests.

Our solids were apple, grapefruit, and potato; whole, halved, quartered, and three-quartered (the three-quartered object was a whole apple, grapefruit, or potato with a quarter-section removed). Our liquid sample of proportion was a glass cylinder that was filled, half-filled, quarter-filled, or three-quarters-filled with tinted water.

Every proportion was used as the sample—the standard against which other proportions were to be compared. And every proportion was used as one of the two alternatives to be chosen as a match for the sample. For example, a quartered section of apple was the sample; a three-quartered apple and a quartered section of apple were choice alternatives. This test was followed by, perhaps, a whole potato as sample with a half potato and whole potato as choice alternatives.

After the series of solids, Sarah and the juveniles were tested on a series of liquids, using a half-filled cylinder as the sample, filled and half-filled cylinders as choice alternatives, and so on. Sarah and the juveniles passed the tests handily. We cannot be too excited about these results, however; true, these are proportions, from our point of view, but they are clearly look-alikes as well. An apple looks like an apple, quite as half a potato looks like a halved potato, and a quarter-filled cylinder looks like another, similarly filled. There are no surprises. As yet.

In the next series, the animals were required to match a proportion of a liquid with a like proportion of a solid. That is, they were given a half-filled glass cylinder of liquid as sample; a halved apple and a three-quartered apple as choice alternatives. They had now to compare apples and colored water on the basis of proportion only. It would hardly have surprised us if none of the animals managed to pass these tests.

As usual, all the samples, as well as alternatives, were rotated so that the apes experienced all combinations of object and liquid proportions. In addition, all the fruit alternatives were equated for size, so that one-quarter of an apple, for instance, was about the size of a halved apple. In other words, we kept size constant by using small apples as exemplars for whole apples and large apples to provide us with quarter sections.

Sarah parted company with the juveniles on this test series. Little wonder that the juveniles failed to pass; but, to our surprise, Sarah not only passed these tests, she did so from the first problem and continued to perform correctly on the first eight trials, and with remarkable ease. It is not possible to solve problems of comparative proportion on the basis of a look-alike strategy, of course.

On what basis, then, does one resolve test problems that require us to compare the proportions of unlike objects? In order to judge the relation of parts to the whole, we must first be able to reconstruct the whole when shown only some portion (or proportion) of it. To match a half-filled cylinder and a halved apple, we must reconstruct a filled cylinder, reconstruct a whole apple, then judge the half-to-whole cylinders as being equivalent to the half-to-whole apples.

If this is the kind of solution used to solve problems that require

A

B

FIGURE 15 A: *Here Sarah confronts the difficult version of the proportions problem that non-language-trained chimpanzees cannot do. The sample is a glass cylinder half-filled with tinted water. The two alternatives are the portions—one-half and three-quarters—of a wooden disc.*

FIGURE 15 B: *Sarah solves the proportions problem by placing the one-half disc on the paper towel beside the half-filled sample.*

us to compare the like proportions of unlike objects, is one not making an analogy? And if Sarah is solving the problems in the same way, does it not mean she might be capable of analogical reasoning? We had never administered tests to Sarah that required her to judge something like the following: a half-filled bottle is related to a filled one in the same way that half-an-apple is related to its whole. Yet, she had to be implicitly reproducing such an analogy in order to equate the halves, quarters, wholes, and so on of disparate objects.

What makes Sarah's success on these tests particularly surprising is that whole versions of the objects were not present on most of the tests (except on occasion when a whole object was either in the sample or alternative position). She had to reproduce from memory the representation of "whole," quite as, earlier, she had to describe an apple from memory when given the little blue triangle. We decided to test Sarah on a variety of problems in analogical reasoning, and, as we shall see, her ability to handle such problems went from the implicit to the explicit.

Reasoning by Analogy

In 1925, Wolfgang Köhler wrote "Mentality of Apes," and in 1929, Maier's "Tests for Reasoning in Rats" made its debut. Both papers provoked vigorous argumentation and disagreement, since many psychologists insisted that what was called reasoning was no more than a series of associations. One major difficulty for experimenters attempting to study reasoning in nonhumans is that the most clear demonstration of reasoning is found in deductive reasoning. Measuring this kind of reasoning requires that the subject understand language.

But reasoning is not a homogenous category, and some forms, in particular reasoning by analogy, are not dependent on language. If analogical reasoning problems, used for years in measuring human intelligence, are adapted with some care, they can also measure the reasoning capacity of apes.

To solve an analogy correctly, an ape must judge the sameness (or equivalence) between sets of relationships, as in the following: "Is the relation $\frac{A}{A_1}$ the same as the relation $\frac{B}{B_1}$?" It must judge sets or relationships with objects as examples, as in the following: "Is the relation between one banana and another the same as that between one apple and another?" The answer would be yes, because the two bananas exemplify the relation of sameness, which also characterizes the relation between the two apples. For the question, Is the relation of a half-banana to its whole the same as the relation of a half-apple to its whole? again, the answer would be yes, because the proportion of one-half to a whole banana is the same as the proportion of one-half to a whole apple. These examples of analogical reasoning can be solved on a purely visual basis. When stated in propositional terms, the problems sound murky, but the visual representations of these analogies are very clear, can be compared quickly, and are not in the least ambiguous.

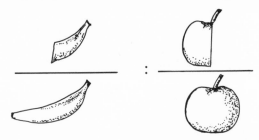

Analogical problems can also be presented in a completion form that asks, $\dfrac{A}{A_1}$:same: $\dfrac{B}{?}$ (the relation of one banana to another is the same as the relation of one apple to what?), or, $\dfrac{A}{A_1}$:same: $\dfrac{?}{B_1}$ (a half-banana is to a whole one as what is to a whole apple). Because Sarah had a history of the proficient use of same and different judgments, we tested her on analogical problems of a wide variety. They were divided into two general sorts of analogies, figural and functional.

The figural analogies consisted of seventy-two geometric forms that were cut from colored construction paper and included all the possible combinations of the following: three shapes, three sizes, four colors, with/without a black dot. For example, $\dfrac{A}{A_1}$ consisted of a small, red triangle over a large, red triangle, while $\dfrac{B}{B_1}$ consisted of a small, green square over a large, green square. In spite of the sensory difference in shape and color, the two sides of the analogy share a common relation: size. In another example, $\dfrac{A}{A_1}$ consisted of a circle over a circle with a dot in it, while $\dfrac{B}{B_1}$ consisted of a triangle over a triangle with a dot in it. In spite of the sensory difference, the two sides of the analogy share a common relation: a transition from undotted to dotted. All other examples were of the same kind. The figural analogies, therefore, were strictly visual, requiring nothing more complex than a sensory solution.

In contrast, the functional analogies required a more complex solution. Sarah had to find the relationship between sets of a variety of household items, such as padlocks, papers, tubes of glue, crayons, knives, paintbrushes, and so on. The relationship between lock/key and can/can opener cannot be judged simply by looking at the visual characteristics of the item; we must know something about their functions. Sarah performed both the figural and the functional analogies at the same level, about 80 percent correct—what might be considered a B or B+ student.

It is not a surprise to demonstrate that very few species can make a same/different judgment. Similarity requires only a reaction to appearance, to how objects look (smell, feel, etc.) compared to one another. No doubt all creatures respond to the similarities among objects. But same/different is an abstract judgment, available only to primates. Primates can use labels, a device that further rescues them from the dependence on appearance alone in making a sense of the world in which they move. We can make judgments with regard not only to colors, but to trees, people, sentences, and ideas.

We are interested in same/different because we believe that the ability to make such a judgment is a prerequisite for language. In order to teach language to an animal, we must first establish that the animal can understand and judge the relation between relations. Only when an animal demonstrates that it recognizes that the relation between two apples is the same as the relation between two elephants, can we hope to proceed to constructions that express real-world relationships, such as: Sarah bites Ann; red is the color of apple; and I want a banana.

When the child, in pointing to a furry creature, says, "rabbit name of that"; in pointing to a red object, "apple name of that"; and in declaring "that not Janet! Mary her name," she is using "name of" in identical ways. For the child sees that the relation between rabbit and "rabbit" (*name of* 1) is the same as that between apple and "apple" (*name of* 2) and that both of them are the same as the relation between Mary and "Mary" (*name of* 3). We are not surprised.

Yet, we should be impressed. In recognizing that the three instances of "name of" are the same, the child has made a judgment concerning the relation between relations, and thereby opens the gateway to language.

Chapter Three

■

Does the Ape Believe You Have Intentions?

■ Children can tell at an early age when someone has hurt them intentionally. They react in one way if they believe the hurt to be accidental but in quite another way if they think the hurt delivered on purpose. Not only do children act intentionally, they think others do, too, distinguishing nicely between truth and lies. They shift their strategy in dealing with a liar. If someone tells the child a series of lies, then abruptly shifts to the truth, the child catches on and shifts behavior accordingly.

Intention is indeed a central concept. When dealing with others in daily life, all of us in fact rely heavily on this notion. We recurrently ask ourselves, What does he really want? Can I trust him? Is she telling me the truth? Does she like me? and so on. Nor do we confine the attribution of intention to our own species. When confronted by a strange animal, we wonder, Is that dog going to bite, or is it just bluffing? Can I trust that wagging tail . . . or is he just waiting for me to turn my back before attacking me? What are his real intentions?

In contradistinction to this widespread popular use of intention, American psychologists (or at least the behaviorists among them) advise us that intention is not only a vague notion, but a bogus one. There is in fact no such thing as intention, the behaviorist assures us; our belief in it is entirely a self-deception.

Quite probably, behaviorists have fallen into the luxury of self-deception on this issue, confusing laboratories with life. In the laboratory, one can easily dispense with intentions. In testing the human subject, instructions can eliminate lying (since human subjects are inclined to tell the truth), so that we do not have to concern ourselves with intentions. And in testing some animals, starvation substitutes for instructions (eliminating any need to speculate about intentions in these experimental situations). After being deprived of food for twenty-four hours (or being maintained at 80 percent of normal body weight), the pigeon or rat whose food is finally restored is likely to do nothing more than eat heartily.

We cannot, however, instruct the chimpanzee (so as to rule out lying); nor is it advisable to transform the chimpanzee by starvation into a mere eating machine. In testing chimpanzees, therefore, we need to retain the notion of intention.

Even if we were to accept, tentatively, the behaviorist's argument that there is no such thing as intention, we could not summarily abandon it. Is it not a striking fact that people everywhere believe in the concept and use it recurrently in explaining their own behavior, as well as the behavior of others? Dismissing the notion does not inform us if humans are alone in believing in intentions. Could some other species—the chimpanzee or even some nonprimates—hold the belief? The topic of intention should not be dismissed but studied.

Indeed, intention needs to be studied on at least two levels: first, which species have intentions? And second, which believe that others have intentions? By definition, lying is intentional falsehood. If someone tells you a lie, you would not believe he did so accidentally. He might bump into you by accident, but lying is often premeditated and always intentional, as are all forms of deception.

Do You Think Sarah Has Intentions?

To find out whether or not animals can lie, we designed a simple test using the four young chimpanzees. It gave them the opportunity to tell the truth in some situations (to lie in others); to recognize and distinguish between truth and lies; and to do all this flexibly—as we humans do. We divided a room into two compartments, separated by a mesh partition, and placed two containers on one side. We put fruit under one of the containers as the chimpanzee watched, then carried the animal to the other side of the partition. Although the ape knew which container was baited, the partition prevented him from reaching it.

On the side of the partition with the containers, across from the ape, we stationed a trainer who literally did not know which container was baited. If he was able to tell, from the animal's behavior, which container was baited, being a cooperative fellow he shared the food with the ape. These friendly trials were followed by others in which an equally uninformed, but *unfriendly*, trainer (wearing dark glasses, a bandit's mask, etc.) replaced the cooperative one. He, too, watched the animal; but he, upon correctly determining which container was baited, kept the food for himself, eating it greedily in the ape's presence.

We then turned the experiment topsy-turvy, placing the chimpanzee in the trainer's position, and vice versa. Now, it was the *ape* who did not know which container was baited. It was now the trainer who tried to inform the ape. And, while the cooperative trainer pointed to the baited container, the unfriendly trainer pointed to the *unbaited* one.

Did these experiences lead the chimpanzee to lie? In these experiments, the chimpanzees showed two forms of lying: full-fledged lying and a precursor form. Only some of the animals became full-fledged liars. At the beginning of the experiment, every animal showed a number of natural, involuntary responses to the baited container. These responses were not done for the benefit of the trainer; they simply reflected the chimpanzee's knowledge concerning the location of the food. The animal either looked at the container, lined up its body with it, moved as close to it as the mesh partition would allow, rocked back and forth accelerating as the trainer approached the baited container or decelerating as he moved away, froze when the trainer approached the baited container, or, almost always, did some combination of these.

These involuntary or unintentional behaviors are of utmost importance when we ask whether the animal lied, because the first form of lying consists of the suppression of these involuntary movements. And when in the presence of the hostile trainer, all four animals did, in fact, suppress these involuntary movements. At the same time, they continued to behave normally in the presence of the friendly trainer.

While suppression was shown by all four animals, only two of them managed to develop actual lying by misdirecting the hostile trainer to the empty container. One animal produced lies of this kind but did not comprehend them. Although she directed the unfriendly trainer to the empty container, she did not understand when the unfriendly trainer lied by directing her toward the empty container. When the trainer pointed to it, the chimpanzee could not suppress her tendency to accept his direction, continuing to check the container even though it was always empty. Another animal did understand the unfriendly trainer and refused to accept the hostile trainer's directive, consistently choosing, instead, the other container. This animal, however, was not itself capable of misdirecting, that is, of telling lies to the hostile trainer.

Only one of the four animals not only produced but also comprehended outright lies by both *avoiding* the hostile trainer's directives and deliberately *misdirecting* the hostile trainer. At the same time, this chimpanzee continued to correctly inform the friendly trainer and to accept information the friendly trainer gave him. While not all animals told lies, those who did, first went through the stage of suppressing their natural (or involuntary) behavior before adopting their lying ways.

Intentional acts undergo an interesting transition. Behaviors that can be used intentionally are not fully developed at birth. Initially, they are elicited (stimulated by specific, outside events), but eventually they end up as voluntary (activated by the internal world of wants and desires).

In the child, these transitions are quite striking. An eleven-month-old grandson greatly enjoyed the game of imitation to which he had been introduced. Whenever the infant made a body movement, the grandmother immediately repeated it, provoking squirming and laughter in the child. At first, the game included only a small number of facial expressions (since, at the age of eleven months, children have a limited variety of spontaneous acts); but as time went on, the repertoire increased greatly.

One day, the child surprised his grandmother in the midst of the game, by introducing a new "face"—he shook his head from side to side, accompanying the shake with a deep grimace of disgust, as though he had just eaten a distasteful item. This act had first been elicited some days earlier by the taste of an olive pit soaked in vodka. Suddenly the infant was able to produce the act voluntarily, without the aid of the martini olive.

The transition from reflexive to voluntary behavior is most complete in primates, though there are noteworthy exceptions, not only in the ape but in our own species as well. We think of our bodies as being unresistingly available for intentional use, but we forget such acts as, for example, contraction of the pupil, erection of the penis, and beating of the heart—acts we *cannot* use voluntarily to express intentions. Such involuntary actions are far more numerous in the ape, not being confined to a few vestigial reflexes (as in man) but involving whole sensory-motor systems. So, in the ape, facial expression and vocalization (cries) remain mostly under reflexive control, elicited by external events rather than being produced by internal desires. The transition from reflexive to voluntary behavior that has affected every major sensory-motor system in man is, in the ape, limited to body postures and gestures of the limbs.

If a chimpanzee were to communicate intentionally with another, he could not do so by either, say, smiling or shouting. He can only stand in a certain way and, perhaps, gesture. However, the gesture of pointing is not natural to the chimpanzee, so when it developed in the course of the experiment, it was a distinct surprise for us. Its emergence as an intentional act, however, is perfectly compatible with the composition of the chimpanzee's sensory-motor system, which provides for some limited number of gestures as expressions of intention.

Most species do not point—or even understand its meaning. They look at the end of a finger rather than beyond it. Chimpanzees at least understand pointing; they look at the object lying in the direction signaled. The most interesting outcome of the deceit experiments lay in the development of pointing. Each animal developed a unique, personal choreography that preceded the gesture (see illustration 16 A-F).

In the first 123 trials, Jessie, a bright young female, behaved like the other chimpanzees, moving toward the baited container and looking at it. On trial number 124, she somersaulted toward the baited container, landing on her stomach with her arm outstretched in its direction. Over the next several trials, she gradually stopped somersaulting. But she continued to stretch out her arm, pointing in the direction of the container.

The animals occasionally looked at both the correct and incorrect containers, so that looking was only a partly reliable indicator, as were all the other involuntary responses. But this was not true with pointing. Once this gesture was adopted by the animals, it became their most reliable response. Moreover, the two animals that lied used pointing to misdirect the hostile trainer. These animals underwent a transition; they emerged from their natural, involuntary looking and orienting toward the baited container, to the suppression of both behaviors, and on to the development of pointing. Once they became accustomed to pointing, the apes directed

FIGURE 16 A-F: *In 1975 we acquired four African-born infants. These animals were not taught Sarah's language, but were tested on other problems. The following six pictures show the development of "lying" in Sadie, the oldest animal in this group. In A, the trainer shows Sadie the container in which he places some fruit. B is the unfriendly trainer, in characteristic costume. He always takes the food for himself rather than sharing it. In C and D, Sadie points—an unnatural act for a chimpanzee. Sadie points with her foot (others used their arms) at the empty container, misdirecting the unfriendly trainer. In E and F, Sadie reacts to the unfriendly trainer's choice. She observes him closely, and as soon as he makes his incorrect choice, her head snaps abruptly in the direction of the container she knows to hold the fruit and she stares at it. In order to lie, it is necessary for the animals to first suppress their tendency to glance at the baited container. All four animals learned to suppress their responses when communicating with the unfriendly trainer.* PHOTOGRAPHS BY PAUL FUSCO.

both the friendly and the unfriendly trainers by pointing—the friendly toward the baited container, the unfriendly toward the empty one.

In Sarah's language lessons, we always asked whether her learning was confined to the training examples, or if it could be applied to new cases. We asked Sadie, our most successful liar, the same question. In one test, instead of leaving the containers across from one another on the floor, we suspended them, one above the other, from the ceiling; she responded appropriately, pointing to the baited container when with her friendly trainer, and to the unbaited one when with her unfriendly trainer. In addition, she lied to the laboratory guard dog, a perfectly innocent German shepherd she nevertheless disliked in the role of a trainer.

Normally, these results would be seen as a substantial transfer, from old cases to new ones—even though confined to the test room. However, when we asked: What would Sadie do in the compound, where the animals roamed freely, foraging on grass and insects, or in the home cage, or the halls of the lab, or in other test rooms? we found that the transfer was really very limited. For pointing never occurred in any location other than that of the original test space.

Even though all four young animals developed pointing on a voluntary basis (we did not train them), none has ever used the gesture outside the test space. Even though the animals could have used pointing in numerous situations (to direct buddies to the location of hidden food in the compound, to deceive rivalrous animals, to indicate to their trainers an object or food they wanted), no animal has ever pointed in "free" space.

This leads us to conclude that while pointing is not a reflexive behavior, it remains conditioned to the test room; it has not fully undergone the kind of transition that frees it of environmental control. Thus, it has not become a completely spontaneous act that the animal can use, voluntarily, to express its intentions.

Nevertheless, let us not forget that pointing developed in the course of the experiment. First, the animals suppressed their normal behavior, behavior that would have indicated the location of the food when in the presence of the hostile trainer. One might have supposed that next they would have used these normal behaviors to direct the hostile trainer to the unbaited container. They would, then, already have been lying. Surprisingly enough, rather than doing so, they completely suppressed these normal behaviors in the presence of both friendly and unfriendly trainers. Instead, they developed an entirely new way of directing the trainers, pointing to tell the truth to the friendly trainer and to lie to the unfriendly one. Clear evidence for intention!

Does Sarah Think You Have Intentions?

We now come to the second level of intentional theory, which is concerned not with who has intentions, but with who thinks that others have them. Does the chimpanzee merely have intentions, or does it reach the second level and attribute intentions to others? To answer this question, we showed Sarah a number of videotapes in which a human actor confronted situations that we would clearly identify as problems. For instance, they depicted a human actor in a cagelike setting, attracted by food (a bunch of bananas) that was out of reach. In one, the actor was shown jumping up and down in a futile attempt to reach bananas suspended overhead; in another, stretching to reach bananas outside the cage on the ground; or, pushing a box that impeded his path; and last, pushing a box that not only impeded his path but also was laden with heavy cement blocks. These videotaped cases were comparable to actual problems that Köhler gave his apes to solve, except that Sarah was not asked to solve these as her very own problems. She was being asked how her *trainer* could solve these problems when confronted with them. Videotapes could be used with Sarah, in part, because of her extensive experience with commercial TV.

Sarah was shown each of the brief, twenty-second videotapes in turn, and the last few seconds of each problem were placed on hold. Then she was given two large still photographs, one showing the correct solution, the other not. For example, the actor was shown stepping up onto a chair, reaching out with a stick, and either pushing aside the box or removing cement blocks from the box. The photographs, placed in a large brown envelope, were given to her by a trainer, who then left the room. Sarah opened the envelope, made her choice, placing the photograph in a designated location, and then rang her bell, summoning the trainer. The trainer graded her choice by saying good or bad in a tone of voice one would use with a child.

In addition to the four videotapes described above, we tested Sarah on four other cases that extended the notion of problem beyond that of inaccessible food. She was shown a human actor in several situations: struggling to escape from a locked cage; shivering and clasping his arms to his chest as he glanced wryly at an unlighted heater; attempting to play an unplugged phonograph; and, finally, trying to wash down a dirty floor using an unattached hose. The alternatives we offered in this series consisted of objects rather than actions performed by her trainer, for example, a key, a lighted torch, an attached hose, a plugged-in cord. In addition, we ran a second series with refined alternatives, for example, an intact key, a twisted or broken one; a lighted torch, one that was burned out or not yet lit; and so forth.

Sarah made only two errors on this series of twelve problems. She failed to select the photograph showing the removal of the cement blocks as the solution in the first series, and she failed to avoid the twisted key. The latter was probably an artifical effect, since the photograph did not clearly show the condition of the key; but the former was not. In fact, Köhler's apes had difficulty with the same problem—and for a very good reason. The chimpanzee, because of its great strength, hardly needs to remove the blocks in order to lighten the weight of a box.

A simple interpretation of Sarah's success is that she can consistently identify the *solution* because, in the first place, she can recognize the *problem*. A problem is, of course, not represented in the videotape. The videotape shows a sequence of physical situations, such as, for example, a man jumping up and down, a bunch of bananas overhead, a cage.

Problems come in countless forms: a car that will not start, a cake that will not rise, glue that will not stick, a recalcitrant spouse (one who will not be persuaded of the soundness of your argument), and the like. While the range of a chimpanzee's problems does not begin to equal the human's, in neither case can the problems be defined on the basis of mere physical appearance.

In order to see the videotape as representing a problem, we must interpret the tape (read it) in a particular way. For example, the man jumping is doing so because he has an intention, as yet unfulfilled, to reach the bananas. A problem can only be interpreted as an unfulfilled intention; there is no way in which to represent a problem as a purely physical phenomenon. It is also important to keep in mind that while identifying a problem in the videotape requires reading it in a particular way, there is no obligation to read it so. That is, the videotape shows a simple sequence of physical events; we can look at it without seeing a problem.

This fact was dramatized forcefully for us by a group of 3½-year-old normal children. When shown Sarah's tapes (and others especially adapted for them, such as a young child jumping up and down attempting to reach cookies on top of the refrigerator, etc.), about half the children chose, not photographs depicting "solutions," but photographs that matched some salient physical object in the videotape. Some children, when shown the videotape of a child attempting to reach cookies located on top of the refrigerator, will not choose the chair, the solution to the problem. They will choose the photograph of a refrigerator, which matches the salient object (refrigerator) shown in the videotape; the chair, of course, does not appear in the tape at all. This kind of choice contrasts with those Sarah made and suggests that the child, unlike Sarah, has failed to interpret the videotape—does not attribute intention to the actor but views the film simply as a sequence of *physical events*. It is interesting to note that 50 percent of the 3½-year-olds we tested responded at a sensory level, in terms of appearance, choosing the alternative that looked like something in the videotape. Sarah, 16 years old when these tests were run, no longer responded at a sensory level.

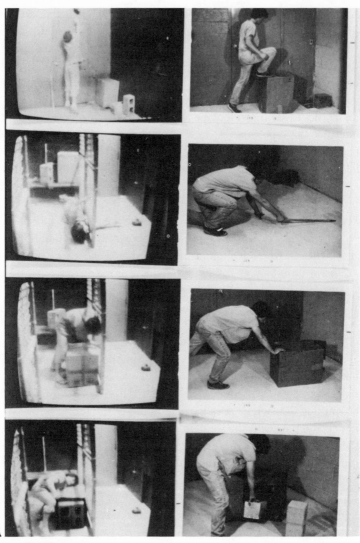

A

FIGURE 17 A and 17 B: *Videotape problems. In* 17-A *the left column shows the terminal image on each of the four videotape problems given Sarah. From top to bottom they show the actor reaching for bananas overhead, for bananas out of reach on the horizontal, for bananas impeded by a box, and for bananas impeded by a box filled with cement blocks. Corresponding to each problem is a photograph showing the solution: the trainer stepping on a chair, reaching out with a stick, pushing aside the box, and removing the cement blocks. In* 17-B *the left column shows the terminal image on more complex videotape problems*

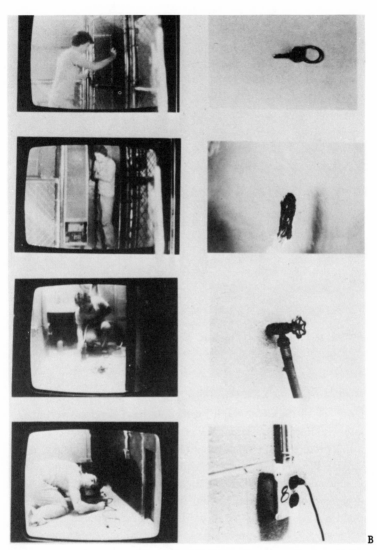

B

given Sarah. From top to bottom they show the actor locked in the cage, shivering by an unlighted heater, unable to rinse a dirty floor, and unable to hear an unplugged phonograph. Solutions to each are shown in the right-hand column: a key, a lighted paper wick, a connected hose, and a plugged-in cord. In every case except one, Sarah was able to choose the photograph that would correctly solve these problems. Sarah's correct choices demonstrate that she interpreted the filmed sequences as depicting problems faced by the actor.

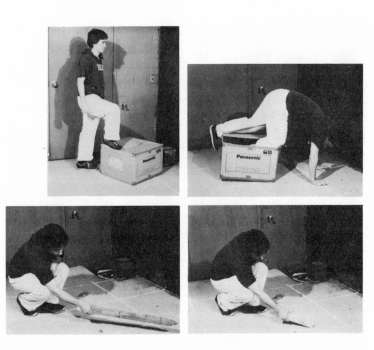

FIGURE 18: *In this series of pictures, Keith, an actor whom Sarah liked, demonstrates both good and bad solutions to the inaccessible food problems. Sarah*

When Sarah examined the videotape and, after a period of study, chose one of the photographs, she could have been answering any of several implicit questions: for example, What would the actor do in this predicament? What should he do? What would I like to see him do? We could ask these questions explicitly of a person—and he or she could distinguish among them (an extremely interesting facet of human intelligence). This facet informs us that people can respond not only to the actual world, but also to potential worlds, better worlds, worlds preferable to the actual one experienced. We have no evidence that the ape or any other species can respond with this complexity.

To find out which of the several possible questions Sarah might have been asking herself, we refilmed some of the videotapes using two different actors: one she liked and another she did not like. We

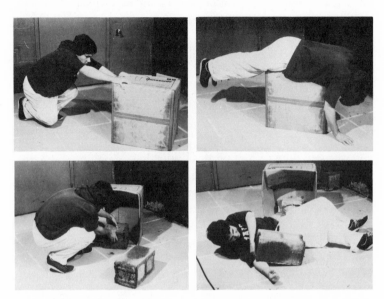

chose all the good solutions for the actor she liked, bad ones for the actor she disliked.

also filmed two sets of solutions: a success and a minor catastrophe, with each role performed by each actor. The "good" set showed appropriate solutions to the problems, for example, stepping up onto a chair, reaching out with a stick, and so forth. The "bad" set showed calamities (accidents that occurred in the course of solving the problem), for example, the actor on the floor with the cement blocks strewn over him, falling through the chair, stumbling over the box, and the like.

Sarah was tested as before, though now she was given three alternatives on each trial: one good, one bad, and one both irrelevant and bad. She was not told which photograph was correct but was praised for all her choices.

Her choices closely reflected her attitude toward the actor. For her preferred actor, she chose almost entirely correct solutions;

whereas for her nemesis, she singled out photographs depicting the minor catastrophes. In addition, she almost never made an irrelevant choice, that is, she always chose a photograph that was appropriate to the problem.

In another series of tests, we asked whether intention was the only state the ape attributes to others. The list of mental states that people can attribute to others is astoundingly long: for example, hoping, believing, knowing, doubting, guessing, and trusting. And we wondered if the ape stops at intention. Can it make other social attributions? We singled out the difference between guessing and knowing and tested Sarah on this distinction. We showed her videotapes in which, on some trials, the actor closely watched the baiting of one of four opaque containers (and thus would know which was baited), and on other trials chanced to be looking away during the baiting (and thus could only *guess* which was baited). We gave her trials of these two kinds for an actor she liked and for one she disliked, and then required her to make a choice between photographs of the actor choosing either the baited or the unbaited container. Sarah gave no evidence of observing the guess/know distinction. As in the previous study, she chose in keeping with her attitudes toward the actors. If she liked the actor, she picked the photographs of him choosing the container that had been baited (on that trial); if she disliked him, the photographs of him choosing the container that had not been baited (on that trial). The outcome of both these series suggested the kind of question Sarah might have been addressing. It seemed much closer to What would *I like* to see him do? than What would or should he do? This outcome reminds us of her unwillingness to be any but the donor in the language lessons concerned with the giving exchange. She had shrieked and shaken the bars of her cage when the written instruction directed that food be given to Gussie. In this experiment, the adult Sarah shows as much emotion about outcomes that concern others as the young Sarah showed about outcomes that concerned her only.

Can we be certain that Sarah did not choose alternatives on some basis other than that of attributing intention? Here are three alternative explanations, each of which can be discounted. First, we might suggest that Sarah, too, in her own way chose physically matching alternatives. To rule out this possibility, we made certain that all four objects—chair, stick, box, and cement blocks—appeared in every film, particularly in the terminal image of every film; this served to prevent Sarah from simply finding an association between a particular film and a particular photograph.

Second, we might say that Sarah did not choose matching objects per se but those that were associated with something in the film, for example, the key in one problem because it goes with lock, and so on. Of course, in order to choose the key, Sarah must know that keys open locks (as lighted torches do not), but is that sufficient? Suppose Sarah were given the same film minus a struggling actor— with only a locked door or with an actor struggling to emerge from an unlocked door—or a film with both the actor and a padlocked door, but with the actor skipping rope. Would Sarah still choose the key?

Of these three cases, we have tested the last one. In videotapes showing the actor shivering and the door behind him both closed and padlocked, Sarah did not choose the key as a solution. This suggests that though an association (of some kind) between key and lock is a necessary condition for choosing the key, it is not suffi- cient. Rather, the actor must demonstrate a specific action, such as struggling to open the door, for Sarah to *choose* the key. In the film showing a shivering actor glancing at an unlighted heater, Sarah chose the lighted wick (rather than a yet-to-be-lighted or a burned- out wick) as the solution. Yet all three conditions of the wick had been about equally associated with the heater. Even more decisive evidence against a simple associationistic account comes from her choices for the actor she dislikes—she had never seen him fall through a chair or strewn with cement blocks, or tripping over a box, and so on. We can conclude that Sarah makes choices on grounds that are far deeper than a superficial "X is associated with Y."

Third, a critic might suggest that Sarah is not choosing outcomes for others, but for herself. She is indicating what *she* would do with the problem. That seems reasonable and appealing, but her choice of accidents or calamitous outcomes for the trainer she dislikes is not compatible with this option. On the whole, the data are compatible with an attributional interpretation, that is, Sarah attributes intention. Her consistent choice of solutions presupposes the concept of problem, and a problem, as we have seen, cannot be defined on the basis of physical appearance. It depends on perception of a mental condition, that is, the attribution of intention, which is the recognition that an intention is thwarted or unrealized in some way.

Do You Think That Sarah Thinks That Keith Has Intentions?

We have examined intention on two levels: having them and attributing them. There is, in fact, a third level of intention: claiming that others attribute them. That apes have intentions is not questioned; probably most species do. We have just reviewed evidence suggesting that, in addition, chimpanzees may attribute them to us. Can we take the last step and claim the third level, the attribution of attribution? Does Sarah think we attribute intentions to her?

Human children can say of another child who is reaching for a ball that the child wants the ball, or of a child extending her hand in the direction of a lollipop that the child wants the lollipop. But they might have some difficulty when shown the picture of a friend looking at a child reaching for a ball or a lollipop. Would a child predict that his friend would say, "Johnny wants to play with the ball" or "Johnny wants to eat the lollipop"? The first analysis is of simple attribution; the second is of the attribution of an attribution, which is much more complex. We do not yet know at what age the ability to make such attributions is possible in the child.

Let us describe the kind of test that one must pass in order to claim the third level. In this kind of test, Sarah is no longer shown an actor who has a problem; she is shown an actor observing someone else who has a problem, and she must indicate how the observer will interpret the actor's behavior. In short, the observer in the videotape is now playing the role Sarah formerly played—looking at someone who has a problem—and Sarah is shifted upward to the role of the experimenter.

For example, Sarah is shown Gussie watching Keith jump up and down below a bunch of fruit. The (embedded) videotape that Gussie is watching is stopped or put on hold, and she is shown being offered several photographs of Keith stepping up onto a chair, reaching up with a stick, and so on. Sarah's own videotape is stopped, placed on hold, showing the scene of Gussie confronting the several photographs, and Sarah is presented with several photographs—one showing Gussie selecting a picture in which Keith steps up onto a chair, or selecting a picture of Keith reaching out with a stick, and so on. Sarah's task is to choose the photograph depicting the choice that *she* believes Gussie will make. To perform correctly, Sarah must attribute to Gussie the capacity of attributing intention to Keith, that is, she must attribute an attribution. Human adults can pass this test; but apes cannot. Apes have intentions and probably attribute intentions but the attribution of attribution is restricted to humans.

Chapter Four   ∎

The Ape
and Its
Physical World:
Action and
Conservation

∎ In consistently choosing correct solutions to Keith's problems in the videotapes, Sarah showed not only that she attributed intentions to Keith, but also that she had a good understanding of the problems that he confronted. She knew that when food was out of reach overhead, one stepped on a chair rather than pushed aside a box. And she knew that when food lay beyond the cage, one did not step on a chair but reached outside the cage with a stick. Sarah's understanding of problems, however, went well beyond the narrow realm of inaccessible food—the kind of understanding that Köhler, many years ago, showed that his apes possessed.

Sarah included, in her understanding of problem, such mechanical difficulties as disconnected electrical plugs, cut hoses, nonfunctioning gas heaters, and the like. For an unlighted heater, she recommended a lighted wick rather than an unlighted one or one that had burned out. For the disconnected phonograph, her solution was a cord plugged into its socket—not a cut cord or a cord that was merely placed in the vicinity of its socket. Her understanding of these problems was so keen that in assigning poor solutions for the trainer she disliked, she not only picked catastrophes for him, but catastrophes that were entirely appropriate to each situation that he confronted.

For example, when food was out of reach overhead, she did not choose to have him fall through a box or sprawl on the floor covered with bricks. She chose instead the photo that depicted him stepping right through a flimsy chair seat. Sarah acquired this kind of understanding of the notion of problem through observational learning alone.

Her day was always very full, and included specific lessons in language on which she received careful, daily drill. However, a very large part of Sarah's day included the cleaning of her cage by her various trainers, meals of various kinds, visits from us and from students (sometimes bringing her favorite foods, such as dill pickles), the playing of records, appropriate television shows, access to her exterior cage and its view of the pond, and so on. Thus, Sarah had ample opportunity for observational learning, even though she was somewhat confined.

Just as the child is drilled in school—taught the multiplication tables, the names of American presidents, the capitals of the world—Sarah was painstakingly drilled in her language lessons: to associate words and exchanges, to order words in her constructions, and to reply to our questions. During the rest of the day, she was left alone, excused from classes and free to play or observe her world at her leisure. She watched the workings of the phonograph, the fine points of cage cleaning, and the procedure for lighting the gas stove. She came to recognize the correct operation of mechanical devices in the same way she came to attribute intentions to others. Without drill. Her analysis of the physical world, like her analysis of her social world, was natural and spontaneous.

Was Sarah's analysis of the physical world really as good as it seemed from the tests? To find out, we might have continued to fill her day with ever more mechanical devices, checking her ability to learn about their function on an observational basis. Or, we might have used Köhler's technique, giving the apes problems—real-life ones, such as silent radios—to see if the apes could resolve them. We chose to do neither but, instead, approached the test of Sarah's ability in a more abstract way. Our approach was analogous to the difference between the ape actually wetting a sponge and its showing that it understands the question What makes a sponge wet? by answering, "Water."

Even the monkey is able to wet a sponge, but we doubt that if asked what makes a sponge wet, the monkey could provide the correct answer. To find out whether or not the chimpanzee could analyze the world in terms of actions such as making a sponge wet, we devised a visual representation of action as a three-termed sequence. At one end was an object in its initial state, such as a whole apple. At the other end was the object in its terminal state, a halved apple; while in between rested a plausible instrument, such as a knife. As you can see, a wealth of actions can be represented in this simple way. In spite of the enormous diversity of actions, they can be represented in the common form: initial state, instrument, final state. Marking, for example, was represented by: a sheet of paper with no marks on it, a pencil, and a sheet of paper that had scribbles on it. Wetting was represented by a dry sponge, a container of water, and a wet sponge.

We chose cutting, wetting, and marking because all the chimpanzees had become familiar with these actions during their casual play. In addition, the three language-trained animals had already been taught plastic words for these actions. Furthermore, the chimpanzees had derived much pleasure from performing these actions. They certainly were never very excited about giving things away, but they threw themselves enthusiastically into cutting an apple with a knife. Marking a clean sheet of paper was so stimulating that throaty hoots (erections in the male) were not uncommon. The greater the pleasure an act held for the chimpanzee, the more swiftly was its plastic name learned.

How were we to find out if an ape that had performed, say, marking could, in addition, tell us what it knew about this action? If all our animals had been language trained, we might have been able to ask them the necessary questions using the plastic words. But the young juveniles were not trained in language. And not all the older animals had the necessary vocabulary to understand questions. Even so, we managed to interrogate all the animals by devising a method of questioning that did not depend on words at all.

The "Still" Representation of Action

At the top of a large green tray, we drew three evenly spaced circles and placed an apple in the first circle, a knife in the second, and the plastic word for question mark in the third. For the juveniles, who had not been taught any plastic words, we left the third circle empty. At the bottom of the tray, we offered three alternatives: a cut apple, a cut orange, and an apple pierced as by a nail. Readers who understand the action will perceive the question and immediately choose the correct answer. Removing the plastic question mark, the interrogative particle, they will replace it with the cut apple. We asked similar questions about the other actions, marking and wetting, by placing appropriate objects in the circles: a blank sheet of paper, a pencil, and a blank or a plastic question mark for the former, and a dry sponge, a bowl of water, and a blank or question mark were offered for the latter. The animals could choose from three alternatives in each of the action problems.

Our present constructions bear a close resemblance to the early language questions we had asked the chimpanzees. The advantage of the new form of interrogation is that we can ask questions of almost any uneducated chimpanzee. We do not first have to teach the animal a language in order to ask it questions. We can determine something of the ape's knowledge concerning its world even though the animal cannot "write" a word.

We have given Sarah several forms of language questions, and in these tests, we were equally able to ask a variety of questions. In asking Sarah about the notions of same and different, we placed the question mark in several places. Now, too, we could leave the mid-

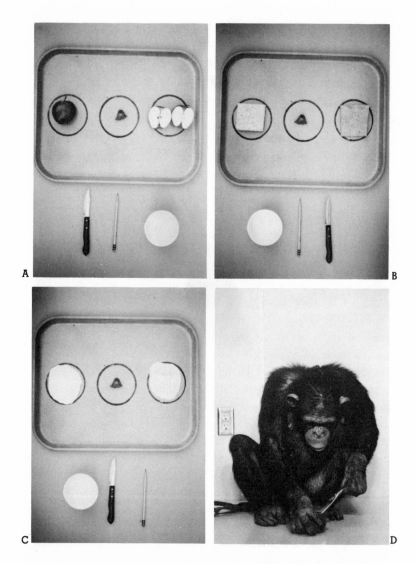

FIGURE 19: A, B, and C are action tests for cutting, wetting, and marking, respectively. The missing item is the instrument. In D, Elizabeth enthusiastically cuts an apple. It was discovered that the more highly the animal enjoyed the action, the more quickly it would be able to understand the related test format.

A   B

FIGURE 20: *An action test on cutting. A shows the problem with its three alternatives, a pencil, a bowl of water and a knife. Sarah is already reaching for the knife. In B, Sarah places the knife in the required circle.* PHOTOGRAPHS BY PAUL FUSCO.

dle circle a blank, asking in effect, What instrument cuts the orange or marks the paper or wets the sponge? In such a case, the alternatives consisted of knives, pencils, and bowls of water—the instruments responsible for the halved fruit, marked paper, and wet sponge.

The chimpanzees that had not been language trained failed every aspect of the action questions, while the language-trained animals—Sarah, Peony, and Elizabeth—not only passed the tests, they replied correctly from the very first test. Another animal that failed was Walnut. A late acquisition from a carnival, where he had been trained by severe punishment, Walnut never engaged in any form of play. He was a deviant chimpanzee in this respect. Though Walnut managed to learn a few words, he failed the action questions as resoundingly as the juveniles.

We had difficulty believing that the untrained animals would fail what the language trained animals could so easily pass. We tried to drill the failures, offering them a repetition of the tests once a month, rewarding them generously when they were correct. But they did not improve. At the end of the four months, they responded at chance level, exactly as they had responded when the tests began. We tried a different approach with Walnut. Since he had never previously engaged in such actions as cutting by himself, we tried to teach him all the actions by the use of passive guidance.

First, to be sure he observed the act properly we cut a banana under his very nose. Then, we placed his enormous hand on a knife and helped him cut through the banana. We assisted him again with wetting and marking. But passive guidance had no effect with Walnut; it helped him as little as monthly drills had aided the juveniles.

Can we explain the success of the language-trained animals on the action tests? Perhaps the language training itself has made the difference. Although Walnut, who failed the tests, received some amount of successful language training, he knew only words. He never reached the stage of combining the words to form constructions, and it may be constructions more than words that make the difference. The animal that has been properly trained understands that each construction stands for a specific condition, such as "red on green" stands for a red object on a green one. Being trained to read written constructions in this way may help the animal to do essentially the same thing with the constructions in the action tests—to read them in a specific way. Let us consider some possible interpretations that could be made of these test constructions.

An actual question that is made up of plastic words—for example, "? apple same banana"—has only one meaning; but the sequence of "apple-knife-apple slices" is subject to countless interpretations. It could mean one-one-two, round-long-sliced, soft-hard-soft, and so on. In order to choose the correct answer, the animal must read the series in only one way: as a representation of action. He must see the construction as an incomplete representation of action, so that he is able to substitute the knife as the instrument in the sequence "apple ? cut apple" and is equally prepared to supply the cut apple in the construction "apple knife ?". It is remarkable that these animals see this whole series as a representation of actions. Actions are, of course, dynamic processes, while these simple constructions are completely static. The videotapes in which Sarah discerned that Keith had a problem at least preserved the motion of real action.

Not so with these tests; they require the animal to "supply" motion to a series of still lifes.

Mere associations could not be responsible for the success of the language-trained animals. In the original test, each object, instrument, and transformed object was changed at every trial. We used apples, cookies, balls, and sponges in the tests of cutting, with paring knives, serrated knives, and dinner knives as instruments and halved apples, cookies, balls, and sponges as the transformed objects. We varied all test items for the cases of wetting and marking as well, never repeating any specific object-instrument pair. Many of the pairs we next introduced were not only novel but anomalous, such as cut ping-pong balls, paper soaked in orange juice, and apples covered with writing. The three chimpanzees performed as well on the anomalous cases as on the conventional ones. They chose the pencil in the presence of an apple covered with writing, the knife for the cut ping-pong ball, and juice for the wet paper.

The animals that were not language trained were given repeated opportunities to pass the test on an associative basis, but they never did so. The four chimpanzees that failed tried to solve the tests in the same way that very young children solved Sarah's video test. When the question involved an apple, the animals chose the red pencil, using the matching strategy that young children used in the video test. Clearly, neither these animals nor the children could be reading the test in terms of a problem, or an action; they were interpreting the constructions on a sensory level only.

The Knife Cuts, but Tape Mends

The human concept of action includes the transformation of an object. We have, in addition, an appreciation for the temporal course of an action: actions involve a beginning and an end. Our tests have consistently been described as having an initial state and a terminal state. In all our tests, we placed the intact apple, the blank sheet of paper, or the dry sponge on the left of the tray. The transformed objects were placed on the right side of the tray.

In other words, we mapped the temporal course of the action onto the spatial sequence, placing cause on the left and effect on the right. Do chimpanzees read the sequences in this manner? Perhaps they simply choose the instrument associated with the final transformation. If so, it would make little difference whether the changed object appeared on the left side of the tray on some trials and the right on others. Currently, we have no evidence that the chimpanzee analyzes action into cause and effect.

But might the chimpanzee be capable of using a left-right order as an equivalent of temporal sequence? Sarah gave us a resounding answer to that question. In the context of play, we acquainted Sarah with three actions: the opposite of cutting, wetting, and marking. She was given the opportunity to erase her markings with a large gum eraser, to dry her wet areas with a cloth, and to mend the cut items with tape. Sarah was now in command of three pairs of acts: one half of the pair damaged an object, while the other half repaired the damage. She could cut and mend, wet and dry, mark and erase.

Sarah had to show us she knew when to cut, when to dry, when to erase, and so on. To do so, we taught her to read the sequences from left to right, giving her "blank sheet ? marked sheet" on one trial and "marked sheet ? blank sheet" on the next. She was offered the same three alternatives (a pencil, an eraser, a bowl of water) on both of these trials, always including one irrelevant alternative. Sarah received numerous sequences of this kind and was informed when she was right and wrong. After this thorough training, Sarah was given a transfer test with sixty new cases. She was never informed if she made an error but was soundly praised for every choice. She was correct on forty of the sixty cases. This is a highly significant outcome. With three alternatives, one could guess correctly one-third of the time; Sarah was correct two-thirds of the time. Such an outcome could occur by chance only one time in a hundred.

Irrelevant Instruments and Actions

Clearly, Sarah had considerable understanding of action—and this on the basis of no specific training. She could be taught to read the order of an action as proceeding from left to right and to choose a knife in the sequence "whole apple halved apple" and tape in the sequence "halved apple whole apple." She picked a pencil in "blank sheet marked sheet" and an eraser in "marked sheet blank sheet." Could she now examine her alternatives and indicate which one was relevant to an action, and which irrelevant? The new tests once again testified to Sarah's considerable understanding of action.

We offered Sarah constructions showing only the initial and the terminal states of an object. She was to choose the instrument that produced the change and to place it in one container. The irrelevant alternative(s), the instrument(s) that played no part in the action, was to be placed in a different container. These tests contained a complex element, however. A sponge, rather than simply being wet, might be both cut and wet in its terminal state. If so, Sarah needed to place both the knife and the liquid in one container, irrelevant items elsewhere. Sarah performed extremely well on this series of rather complicated tests. It was clear that she could properly examine the objects in the first and third circles on her tray, compute the difference between them (cut and wet or marked only), and identify the proper instrument (or instruments) that could produce these changes.

Conservation: Actions That Do Not Change Objects

Sarah's ability to distinguish between relevant and irrelevant instruments suggests rather strongly that she could also determine those properties of an object that have been changed and those that have not. Cutting an apple into four or eight slices may change the number of pieces of apple, but cutting does not change the total amount of apple. Painting a sheet of paper blue changes its color but does not change its size. The instruments we used in the action tests are rather special, for they change some aspects of an object but leave other aspects unchanged. Pencil marks change the surface appearance of a piece of paper but leave the paper's shape, size, odor, and so on unchanged. While we find these facts perfectly obvious, they are less obvious to children.

If we give a child a row of buttons, nicely spaced apart in a long row, then press the line of buttons close together into a shorter row, does the child think that the number of buttons has decreased? Or, if we pour liquid from one glass into a narrower one, so that the level of the liquid is higher in the second glass, will the child believe the amount of liquid has changed? Tests initiated by Piaget, called tests of conservation, have shown that children who are below a certain age do not, in fact, recognize that these changes are only apparent. The children believe that both the changes are substantive—that pressing buttons closer together decreases the total number of buttons, and that pouring water from one glass into a narrower one increases the amount of water.

It seemed to us that Sarah should be examined on these tests in light of her good understanding of the physical world. But we could not get very far on the button test. Sarah was unable to judge whether two rows of buttons were same or different even before we made any changes in their spacing. Sometimes she judged rows equal or same when they were not, then judged rows unequal or different when they were indeed equal. Since she could not judge number in the initial part of the test, how could she judge any transformations we might make in pressing the rows together or in removing an occasional button on tests? Sarah is not incapable of making judgments concerning number, of course. She made correct judgments involving small numbers on earlier tests, but number seems to be a property with little salience for her. For the child, number is salient at a very early age. We dropped the button test in favor of the tests on liquid and solid amounts.

Sarah did not have the least trouble in making correct judgments of same and different to various amounts of both water and clay.

FIGURE 21 A-B: *The match-to-sample test on numbers. In A, Sarah confronts the problem, and in B, she selects, by pushing forward the correct tray, three spools to match the three metal cups of the sample.*

We showed Sarah pairs of glasses filled with either equal or unequal amounts of water and we showed her two pieces of clay that were shaped alike but contained either the same, or different amounts of clay. Now that we had accurate initial judgments from Sarah, we could move to the next step of the test and transform the shape of both the liquid and the clay. Would Sarah now be misled by changes in the appearance of the water? Of the clay?

Sarah watched with considerable fascination as we poured liquid from one glass into another of a different size. Sometimes we showed her two glasses containing the *same* amount of liquid, then poured the liquid from one of them into a narrower one, so the liquid level now looked higher. Sarah compared the original glasses as same, and judged the transformed ones same. The simple action of pouring·did not change the amount of liquid in Sarah's estimation. Sometimes Sarah was shown glasses that clearly had different amounts of liquid, which she compared as "different." The trainer

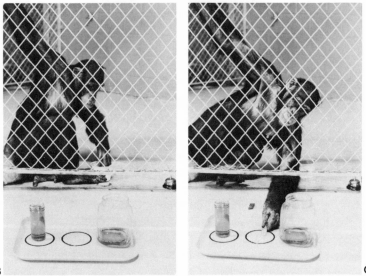

FIGURE 22: *Conservation test on liquid quantity. In A, Sarah watches Keith pour the contents of one full flask into a much larger empty flask. In B, she reaches into the bowl that contains her words "same" and "different" and places the word "same" in the appropriate circle in C. Notice the bell in the right-hand corner of the photo. Sarah uses it to summon her trainer after completing her problem.* PHOTOGRAPHS BY PAUL FUSCO.

would then pour the greater amount of liquid into a wider glass whose liquid level equaled exactly the level of the original. Once again, the equality of the liquid levels was seen as irrelevant—Sarah judged the amounts still different.

The results on tests we gave Sarah involving solid amount, conducted with pieces of clay, were exactly the same. If she judged two pieces of clay the same, no amount of compressing or stretching, actions she observed avidly, led her to consider them different.

But what if Sarah was not really judging the whole transformational process at all? Could she not have been simply saying same in the original judgments, then repeating it? Such a strategy could work well on these tests, but it would have been an odd strategy for Sarah to have adopted, since she was never scolded for her wrong choices. We allayed our suspicions by actually adding or removing some amount of water or clay on each test. During the transformation of liquid, we removed or added some amount of the liquid. And when we did, Sarah judged the amounts different. For Sarah, the action of pouring did not change the amount of a liquid; stretching and compressing were actions that did not change the amount of clay. However, the addition or removal of some quantity of both liquid and solid did constitute a relevant change for her.

Finally, in a kind of ultimate test, we withheld the original test information from Sarah. We did not show her the two identical pieces of clay, but showed her only the result of the transformation—a piece of clay that was more stretched out than the piece beside it. Sarah was quite unable to judge these cases, always responding at random. We, too, have difficulty in making such judgments. In order to judge whether a pair of changed objects are the same, we must see them in their original state, then observe the kind of action that the objects undergo. Some actions transform objects, while other actions have only superficial effects.

When we find that the child and the ape perform tests in the same way, we are certainly intrigued by the similarity. But is the underlying process that each uses the same? It is not particularly easy to identify the process that underlies a performance. We can at least be sure the child and the ape receive the same tests. Normally, conservation tests designed for children are based on speech—out of the question for Sarah, of course. So, we reversed the test. We taught children Sarah's plastic words for same and different and then tested them as we had Sarah. The results were straightforward. They performed Sarah's test with the plastic words in exactly the same way as on their own tests using speech. Young children passed our test on number but failed on tests of amount of both liquid and solid. The older children, of about five or six years of age, passed all the tests. Using plastic words, the children duplicated quite perfectly the results that had previously been obtained when speech was used in these tests of conservation.

Chapter Five

■

Do Apes Cheat on Tests?

■ Some people worry that when apes are presented with problems to solve, rather than actually attempting to solve them, they use social cues from their testers. For example, in the course of being tested on a match-to-sample problem, the animal makes a tentative move toward one of the alternatives. If the alternative is correct, the trainer smiles—and the animal selects it. But if it is incorrect, the trainer frowns, and the animal pauses and turns to the other alternative. In this way, the animal responds correctly on trial after trial, appearing to have solved the problem. In fact, he has been guided by social cues and has not solved the problem at all.

Here we will examine this view of psychological testing: What are the underlying assumptions; How well founded are they; Are they based on test evidence or opinion? The view that animals (and people) inevitably "cheat" when tested, is based on the following assumptions. First, that every experimenter or tester unwittingly provides social cues. The tester does not know what the social cues are—no one does!—and therefor he could not control them even if he wanted to. Social cues emanate from a tester the way smoke emanates from a fire. Further, the individual being tested is drawn, just as unwittingly, to the social cues. He is no more aware of using them than is the tester aware of providing them. Why does he seek social cues? Because his real interest is not in solving the problems, it is assumed, but in being judged right or in being rewarded with praise or tidbits.

Here is a simple procedure that will allow us to test these assumptions. Suppose an ape performs quite accurately on a series of match-to-sample problems using objects. Suppose, in addition, that we use a so-called Clever Hans procedure (a procedure designed to control for social cues) in testing the ape. Either the paper-pencil test or the more traditional double-blind procedure (both of which will be described later) could be used for this purpose. But, if social cues emanate from the tester like smoke from a fire, and if animals are irresistably drawn to them, how can we be confident that the controls are effective?

We give our tester a camera and ask her to photograph ten of her friends and the cars that each owns. The ape, of course, has no knowledge of which car belongs to which owner—only the tester knows that. We use the photographs to make up a special set of match-to-sample problems. The car is the sample in some cases (with the owner and a nonowner as the alternatives), and the owner is the sample in other cases (with his car and someone else's car as the alternatives). The car problems are interspersed with the regular match-to-sample problems, and testing goes on as before. The tester praises all the animals choices.

Suppose we obtain the result that the animal continues to solve the regular problems but performs at chance level on the car problems. What would these results tell us? Well, at the least, that the Clever Hans control is effective. For, even if testers provide social cues inevitably, and animals are drawn to them irresistably, there is, nonetheless, a procedure for eliminating them.

But we can go beyond simply testing the effectiveness of controls against social cues. We can test the accuracy of the assumptions people have made about social cues. To do so, we administer the same test as before—the regular match-to-sample problem, along with the special car problem—and at the same time eliminate all the Clever Hans controls. We neither use two testers, as in the double-blind procedure, nor remove the tester from the room, as in the paper-pencil procedure. Instead, we deliberately use only one tester, an individual who knows the answers to all the problems, and we allow her to remain throughout the test. This informed person presents every problem to the animal, her face only inches from that of the animal. The opportunity for providing and receiving social cues should be at its zenith.

Suppose we obtain the same results with this procedure: the animal passes the regular object-matching problems but continues to perform at chance level on the special car problems. What impact do these results have on assumptions some people have drawn concerning psychological tests? Clearly, either one (or both) of the Clever Hans assumptions is wrong. Either the trainer does not give off social cues, or she gives them off but the animal does not use them; or she does not give them off, and the animal is not intent on being cued. Additional tests can be made to distinguish among these alternatives. We want to make explicit the kinds of assumptions that lead some people to worry about "cheating" in psychological tests.

We gave Sarah a series of formal tests to determine the effect of inadvertent social cues. We started with a test of Sarah's ability to process her language by using an uninformed trainer, one who behaved as though he knew the language, when, in fact, he did not. Because Sarah would not work with any but people familiar to her, it was necessary for the uninformed trainer to spend at least some hours with Sarah. This he did simply by coming to visit and play, without engaging in any language work with her. After it became clear that Sarah and the new trainer could get along reasonably well, he served as the uninformed trainer in one of the tests for the effect of social cues on the outcome of Sarah's performance.

The trainer was able to simulate a speaker, pretending to know

Sarah's language, because of two advantages we gave him. First, we gave him a code consisting of a picture of each word with a corresponding number. For instance, the blue triangle meaning "apple" was number 16, the pink square, her word for banana, was number 3, and in like manner, all the words were assigned numbers. So, while Sarah knew the meaning of each word, the trainer did not, but knew only the number that corresponded to each plastic shape. We used the code (along with his written instructions) at the start of each lesson; as, for instance, "Display, on Sarah's table, a piece of apple along with the words 12, 9, 3, 20, 4, 16, 8, 11." The trainer also used the code to describe the construction or sequence of words that Sarah composed. The code was supplemented by the second advantage, a translator in the hall, to whom the trainer described Sarah's productions through a microphone. The trainer, meanwhile, received the translator's instructions through earphones.

How the system worked can best be illustrated by a sample trial. After setting out the objects required in the written instructions, the trainer waited for Sarah to make her choices among the words and arrange them into a vertical sequence. Then he used the code to describe Sarah's composition: "12, 6, 3, 4." Then he waited for the translator to decode his message and say, "right" or "wrong," the only two words he ever heard over the earphones. If he heard "right," the trainer informed Sarah "that's right," handed her the object he had set out for her, and arranged words and objects for the next trial. If he heard "wrong," Sarah was informed "that's wrong," and the trainer did not hand her the object set out. Once again, arrangements were made to proceed with the next trial.

Did Sarah recognize that the trainer did not know the language? That he was only pretending? We were not able to ask her such a question. If she did suspect the trainer of ignorance, she tolerated his pretence, allowing the lessons to proceed even though her own performance began to deteriorate in peculiar ways.

On the production tests, in which Sarah was required to compose acceptable constructions, Sarah was given a set of eight words—five of which were relevant to the situation. For instance, in three of the production tests, the objects placed before her were, respectively, apple, banana, candy; candy, nut, cracker; cookie, banana, candy. A representative set of the words given to her consisted of "give, Sarah, candy, nut, cookie, John, Debbie, and eat." Sarah had to ask for one of the objects placed before her by constructing a four-word construction using the form "John give X [the name of the object] Sarah." The uninformed trainer wore his name around his neck, like a medallion, as Sarah wore her name—as did the rest of the trainers who had worked with Sarah. We were confident that, although Sarah had not specifically been taught the "dumb" trainer's name, she would make the necessary inference concerning his proper name, and she did. In the three tests of production, she made 2 errors in 16 trials, 6 errors in 20 trials, and 2 in 11 trials.

A second production test involved simple questions of, for example, the following kind: (1) "Red ? apple" (what is the relation between red and apple?). The answer was "color of." (2) "Small ? nut" (what is the relation between small and nut?). The answer was "size of." (3) "Square ? caramel" (what is the relation between square and caramel?). The answer was "shape of." (4) "Orange ? object orange" (what is the relation between orange and the object orange?). The answer was "name of." She was given every alternative with each question, as well as "if-then," an alternative that was irrelevant to any of the questions in the tests. She made three errors in ten trials in this set.

In the comprehension tests, the uninformed trainer placed a number of words on the board in a column, according to the numbers listed on his instruction sheet. He then laid out a set of alternatives, such as a red and a green card, along with a pail and a dish. After Sarah chose an alternative, the trainer broadcast the choice to the hall and, depending on the information from the hall, did or did not reward Sarah's performance. On the first test, three cards (a blue, a yellow, and a green) were placed before her, and Sarah was instructed as follows: "Sarah take blue," "Sarah take yellow," or "Sarah take green." She was correct on 11 of 15 and on 12 of 17 of these tests. On a second type of comprehension lesson, a yellow card and a green card, a plastic cup, a cracker, and a pile of nuts were placed before her on each trial. She was given instructions of the following kind: "John insert yellow if-then Sarah take cracker," "John insert green if-then Sarah take nut," and so on. She was correct on 7 of 10, and on 8 of 11 trials.

In these tests, 58 varieties of sentences were used. Probably a chimpanzee could memorize this many sentences. But the problem with attributing correct performance to memorization in Sarah's case is that by the time she received the social-cues tests, she had experienced somewhat more than 2,600 different sentences. Of course, it was not possible for Sarah to predict which 58 of the 2,600 should be included on these tests. In order for Sarah to have done well, she would have been required to memorize not 58, but all 2,600 sentences. Further, while many of the sentences that appeared in the social-cues tests had been seen or produced by Sarah in earlier tests, about 25 percent of the sentences given her were entirely new; she performed about as well on the new as on the old sentences.

Sarah performed about 10 percent less well than usual on all of these tests when they were conducted by an uninformed trainer. Are we to assume, then, that about 10 percent of her correct answers are due to the unintentional cuing from her informed trainer? There are some other differences in her performance in these tests besides the reduction in accuracy. Sarah regressed. She produced constructions using a procedure that was typical of an earlier period—once again placing correct words on her language board, but in an incorrect order, and then rearranging them. Although she had abandoned this process some ten months earlier, she resumed the practice when being tested by the uninformed trainer. Further, she abandoned her neat, orderly manner of writing, and adopted a sprawling construction that was also characteristic of her "baby talk." These regressions suggest that Sarah was somehow strained or frustrated in communicating with someone who did not know her language but was pretending he did. I wonder how we would manage a conversation with someone who pretended to know English? Probably not well, considering how exhausting we find even simple discussions with foreigners who wish to practice English.

At the end of these tests, we examined the uninformed trainer to see how much he might have learned in the process of administering the language tests to Sarah. In the first test, he was examined on whether or not he could distinguish a sentence from a nonsentence. He could. Of 50 trials, the trainer made only 8 errors, demonstrating that in the course of testing Sarah he had learned to recognize the appearance or structure of Sarah's constructions. In the next test, some 20 of Sarah's words were arranged in 50 new constructions. Twenty-five were nonsensical, and 25 were correct. In this series, 23 of the 25 constructions were entirely new to the trainer. He made 20 errors in 50 trials, indicating that although he could recognize familiar constructions, he had not comprehended the grammar, that is, he could not distinguish a novel, proper construction from one that was improper.

The next test he was given was an exact duplicate of one he had himself given Sarah. A blue, a yellow, and a green card were laid out before him, and he was given the instruction "John take blue," "John take yellow," or "John take green." He made 20 errors in 20 trials—a consistent misassociation between the colors and their names. This suggests that while testing Sarah the trainer made guesses about the meaning of certain words. But the only information he received from the hallway—"right" or "wrong"—was never enough to verify his lucky guesses and eliminate his unlucky ones. Next, he was given a test of Sarah's class-concept words (color of, size of, shape of, name of), using the same test questions that had been given to Sarah. He made 9 errors in 10 trials, again involving a misassociation of words and their referents. While the uninformed trainer had learned some things correctly in the context of the language test, he had mislearned much more.

What, then, is the relationship between the information held by the trainer and the kinds of errors that Sarah made in these tests? Virtually all the uninformed trainer's errors consisted of systematic, consistent misassociations between words and their referents—bad guesses. None of Sarah's errors were of this kind. Sarah did not even try to extract social cues from him on this kind of material, as she did not have to. The material was well-known to her already. In addition, Sarah performed as well on the early sessions of the tests as she did on later sessions. That is, she did as well before the uninformed trainer learned to recognize sentences versus nonsentences as after.

In view of the test results, we hazard the guess that Sarah suffered a 10 percent loss in accuracy—not because she tended to receive better information from her informed trainer, but because there were disruptive aspects to these tests that Sarah had not encountered before. For example, there was an aberrant delay between Sarah's performance and the recognition of her performance. Since assessments of correct and incorrect had to be translated from the hall, the wait was considerable. No doubt young children tested in this fashion would demonstrate a similar decrement in accuracy; for that matter, so might adults.

In general, Sarah's performance in the double-blind tests is highly significant. The probability of choosing four words from eight alternatives and placing them in the correct order by chance is extremely small, 1 in 1690 tries. These tests contrast notably with other tests that have been administered to chimpanzees. Vocabulary tests show the chimpanzee simple items that it must name—an estimate of the animal's response associations to different objects. In our tests, we examined not only vocabulary, but, of more importance, the complex aspects of sentence production and comprehension.

We also designed a simplified paper-and-pencil test that preserved many of the features found in scholastic aptitude tests. In the human test situation, the test allows the instructor to present multiple questions at the same time; the student leaves a mark in a box that indicates his choice of one of the several answers provided. The instructor can, to avoid providing social cues, isolate the student. This language-based test invites application to the language-trained animal, and it is a notable improvement over other widely used testing apparatus. One (designed at the University of Wisconsin) while eliminating social cues, is arduous and time consuming; it is hardly feasible for use in comparative work with human children.

Though this test procedure can be used with almost any kind of

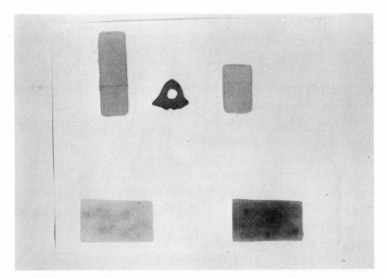

FIGURE 23: *On the top line of the paper-and-pencil test is the question, consisting of two colored forms, one larger than the other, set on either side of Sarah's question mark. Below the question are two answers, Sarah's word for "same" on the right, and her new word "similar" on the left. Sarah answered the question by sticking a piece of tape on one of the alternatives (originally she answered by checking the word with a pencil, but when she became too frisky with the pencil, we substituted the tape). The correct answer here is "similar" because, although the two forms differ in size, they are the same in color and shape.*

test on any subject, we will describe the details of its very effective initial use with Sarah in tests that required her to make judgments of same, similar, and different. We adopted a slight change in procedure when Sarah became enamored of her pencil and started to use it too freely after responding to her question. But the use of a piece of tape to indicate an answer could also be used in human tests without compromising the accuracy of multiple-choice tests. The term "paper-and-pencil test" provides an efficient description for the kind of test we designed and readjusts the animal specialist to the potential of the primate over other types of animals.

We made paper versions of the original language tests that involved the notions of same and different. The words for same and different were used as elements of plastic-word constructions written vertically on a slate, or in hybrid strings of both pictures and words written horizontally on a tabletop: for instance, "apple same apple"—a hybrid construction in which apple was a photograph of an apple. Although Sarah started by replacing her "?" with either "same" or "different," depending on whether the particle sat between two paper clips (same) or a clip and a pencil (different), she eventually progressed to the point where she was judging pairs of constructions as synonymous or not. For instance, she judged the string of plastic words "apple is red ? red color of apple" as "same." She judged "apple is red ? cherry is red" as "different."

In the paper version of these tests, Sarah's task was to mark the one word that accurately described the relationship between two exemplars pictured on the test sheet. Each question consisted of two exemplars and an interrogative particle; each had two possible answers. One example of the five tests that were administered to Sarah in succession is described below. We varied the nature of the exemplars, the set of potential answers, and the number of questions we presented at one time.

The first test showed pictures of familiar toys, generated from a set of thirty colors and forty toys of various sizes. The pictures were traced outlines of the toy, colored by crayon—nothing very elaborate or sophisticated. While a particular picture could appear more than once during the test, each combination of two exemplars appeared only once during an entire experiment. And while three potential answers were used ("same," "similar," "different"), only two appeared on the list accompanying any one question. One answer was correct; the other, incorrect in every test.

The answers were pictures of the plastic words, traced and colored in the manner of the object exemplars. The object exemplars that were to be labeled "same" were completely alike, those to be labeled "similar" were identical on only two of the three dimensions. The "different" exemplars had no resemblance on any dimension. "Similar" was the newly introduced term, and we were interested to see if Sarah could acquire a new word (as well as

FIGURE 24: *This is a continuation of the paper-and-pencil test in which Sarah is asked to compare not differently colored objects as in her original tests, but black letters of the alphabet. Despite these changes, Sarah performed very well, ringing her bell (visible between the two trays) to summon her trainer when she had answered both questions.*

recognize and use her familiar words "same" and "different") in a pictorial mode.

In test series one, each page contained one question, while in test two, two questions were presented on every trial. The test paper was divided by a heavy black line, containing one question with its choice of answers on the space above, another below the line. On test three, four questions were presented on each trial. The individual test questions were drawn as in test two but were cut along the black line, and four individual test papers were offered Sarah on each trial. In test four, Sarah's exemplars were letters of the English alphabet. The letters A, B, C, Y, and Z were stenciled in black ink, and since the letters were always alike in color and size, only same and different were required answers in this test series. In test five, a new set of letters (D, E, H, and I and S) were introduced.

Before starting her paper-and-pencil tests, Sarah was taught to mark her answer with a pencil. We used a combination of modeling and passive guidance techniques to help her along. At first, Sarah cooperated swimmingly, but her enthusiasm for scribbling with the pencil soon overcame the limitations we had set for the test. Although it was often possible to interpret her answer as the word that contained the greatest density of pencil marks, we decided on a cleaner procedure midway through the second test. Sarah was given a piece of masking tape and swiftly trained to put the tape on her word choice. Fortunately, this procedure did not precipitate innovations, that is, Sarah did not tear the tape into smaller bits. The use of the tape was maintained for the remainder of the tests.

Sarah's tests started with one question at a time on each test paper. On every trial, Sarah was given her test paper by the trainer, who then left the room, permitting Sarah to mark her answer choice; whereupon she rang a bell to notify the trainer she had completed that portion of the test series. Her trainer checked her answer, telling her if she was right or wrong. At the end of every test session, Sarah was given yogurt, fruit, or candy.

In spite of the absence of the trainer, Sarah's performance on all five tests was quite good. Her accuracy on making judgments of sameness remained high throughout the series of tests, but on judgments of difference, she was less accurate and fluctuated somewhat from test to test. Her judgments of similarity, which started out as uncertain, came, in time, to be almost as reliable as those for sameness.

Combining the results of the first three tests, we found that Sarah was most accurate on judgments of sameness (eighty-eight percent correct), intermediate on judgments of similarity (seventy-six percent correct), and least accurate on judgments of difference (seventy percent correct). Her performance on sameness resembled her performance with plastic words. In a sense, her better grade on sameness resembles the performance of humans. While we do not make errors on simple problems calling for such judgments, the latency between answering problems calling for either judgment is greater for difference than for sameness. Also, Sarah was more accurate in replying to clear-cut "same" or "different" questions than to questions that required the finer distinctions of same-versus-similar, or similar-versus-different.

The paper-and-pencil test has raised the objection that the trainer knows the answer and could easily give it away by, say, smiling when the answer is "same," frowning when it is "different," and so on. How he would do this (after the first hurdle of the tests is over) when the chimpanzee is given two, and finally four questions at the same time, is not clear—especially since the answers on each test sheet may be any combination of "same" "different," and "similar." Nevertheless, we have anticipated even these critics and conducted a few tests using a trainer who was almost wholly uninformed, that is, did not know Sarah's language or the form of an answer or, indeed, even recognize a question in Sarah's language system. The results we obtained were exactly like those found with the highly informed trainer.

The classical double-blind test calls for two individuals: one who, say, asks the questions without knowing the answer to the questions he asks; and another who records the answer without knowing what the question is. This kind of double ignorance protects against two possible sources of error. If the questioner knows the answers, he might bias the subject's answers by providing facial cues, such as smiling or frowning. If the person recording answers also knows the questions, he might record inaccurately (by cheating or by hearing only what accorded with his own hypothesis of the outcome of the tests).

In the match-to-sample problems on which the juveniles have been tested, two trainers are separated by a barrier. One trainer can see the sample but not the alternatives; the other trainer can see the alternatives but not the sample. This arrangement realizes all the desired features of the double-blind procedure: the questioner does not know the answer, while the recorder does not know the question. We have consistently used this standard format when testing our four juvenile animals to whom we have not applied the language procedures.

There are also numerous informal indications that the apes understood quite well the nature of the tests they were given, and that they were engaged by the problems themselves and not by the facial expressions of their trainers. Furthermore, the apes enjoyed introducing variations to some of their tests. Their innovations were surprising, for they were in no way related to our training procedures. One of the innovations occurred in the context of the yes/no questions, in which the trainer placed, perhaps, a red card on a green card and then asked one of the chimpanzees "Is red on green?" The ape was required to read the question in the plastic language, examine the actual placement of the cards, then answer yes or no, depending upon the agreement between the construction and the actual position of the cards. Sarah did consistently well on this type of question. On some occasions, however, instead of replying "no" (because of her inexplicable general dislike for the negative), she would change the arrangements of the cards in order to answer yes. For instance, if asked, "Is yellow on blue?" when blue was actually on yellow, rather than answer no (yellow is not on blue), she would place the yellow card on the blue card and reply yes (yellow is on blue).

Sarah also engaged in innovations on those rare occasions when not she but the trainer was to be the recipient of some favored food. One of the most amusing replies she gave to "Sarah give apple Mary" was to take her plastic name and stamp it vigorously all over the apple, as if to say "mine, mine, mine". Children not infrequently lick or spit on fruits and candies to discourage others from taking their favorite choices. Instead of claiming the apple and refusing to hand it to Mary, Sarah expressed her dismay in linguistic form, using her name rather than her considerable physical strength to claim the apple.

Sarah had more difficulty learning certain words than others. If she had been relying mainly on social cues, she should have been capable of picking them up equally well no matter what the vocabulary. But this was not true of her performance. The conditional "if-then" and the quantifiers "all, none," and so on, all presented difficulties for Sarah. Some of her difficulties stemmed from our failure to find the ideal exemplars to illustrate her words. When these problems were corrected, Sarah learned the words nicely, indicating it was her ability to understand the tests (rather than her search for social cues) that was responsible for her continued correct performance in transfer tests.

The fact that the animals fail some tests provides a clear proof of the claim we made earlier: some animals do not use social cues at all. Take the number-matching problem for instance. We know the ape cannot do better than match a sample of five items against two alternatives, one with five items versus one with four. The animal that does not use social cues will fail when they are given numbers larger than five for matching. When the animal is given these problems (and others), using a double-blind procedure in which both trainer and animal have incomplete information about the test, of course the animal fails with the numbers beyond five. However, we can give the same animal the same number-matching problems in a test situation where social cues are freely available, and the animal still fails with numbers above five. It was not the effectiveness of the double-blind procedure that prevented the animal from cheating (by using social cues); rather, the animal does not use social cues even when they are available. Even when the animal can solve the problem only by using social cues.

Our formal tests for cheating—the uninformed trainer, the paper-and-pencil, the orthodox double-blind—establish quite clearly that when the chimpanzee succeeds, its performance is based on problem solving and not on the use of social cues. The results of the formal tests are hardly surprising, for they parallel a host of informal results, all of which support the same conclusion. It is clearly demonstrated in Sarah's answer to "Is red on green?" (at a time when red was on blue). Rather than answer no, she replaced the blue card with a green one, answering yes. And it is just as clearly demonstrated when the ungifted Peony, asked about the relations of same and different, placed two like objects together (added the word "same" to them) and then moved the unlike object off to the side, adding the word "different" to it. The trainer could hardly be accused of cueing the animal in these cases, for the trainer was completely surprised by the use of a procedure that was literally invented by the animal itself.

Cheating, when it does occur, is easily detected. But cheating is an uncommon occurrence; for children and chimpanzees are intrinsic problem solvers. Most of them will fail in the course of solving problems rather than use social cues, even when the cues are freely available.

Chapter Six

■

Translating "Pictures" into "Words" and Vice Versa

■ The rat is one of several laboratory species that traditionally has been first deprived of food, then subsequently fed or rewarded for pressing a lever. This procedure is a basic training strategy with animals. One proceeds differently with humans. One simply says, "If you're hungry, put a nickel in the machine, press the lever, and you'll have some food." Obviously, the two approaches are different, but how? What is the difference between training a rat to press a lever and instructing it to do so? But there is no conceivable way, no language to use, to instruct a rat to do the task. Working with the chimpanzee as a test animal, the question can at least be addressed. Clearly, the chimpanzee can be taught a little language and, consequently, instructed. Having taught Sarah the conditional particle (if-then), we could simply write her the instruction, "If Sarah want chocolate, press lever."

This difference between the test subjects is very important, for it tells us that the rat has only one representational system while the chimpanzee, like us, has two. The rat, like the pigeon, is limited by its imaginal system; the ape has both an imaginal and an abstract system. The rat can store its information only with pictures; the chimpanzee, with both pictures and words.

If we merely train the ape to press levers, it will respond as a pigeon does when trained to recognize colors. It will be hampered by physical resemblances and will press only those levers that are similar to the ones on which it was trained. However, if the animal is instructed, it will escape this limitation and will press any object that it will accept as representing a lever. The only essential limit on the ape's performance will be in terms of how general a definition of lever it has.

This theory seems to solve the training/instruction dilemma neatly, but it turns out to be highly specific to the chimpanzee. For, if the same experiment is tried with the child, the result will be quite dramatically different. The child, whether trained or instructed, will respond in the same way. In neither situation will the child be limited by physical resemblance. Even though trained to a specific lever, it will press not only levers that physically resemble those used in training but, in addition, any levers that meet its definition of lever.

So, the difference between training an animal and instructing it depends critically on the animal itself. The rat cannot be instructed at all. The child finds no difference between being trained and being instructed, because it automatically converts the one into the other. It spontaneously translates visual scenes into equivalent sentences, and sentences into equivalent visual scenes. Once the child translates training into instruction, he or she is no longer bound by the limitations of physical resemblance. The chimpanzee is evidently incapable of these spontaneous translations. Perhaps it can convert such instructions as are given to it into visual scenes or images, but it is incapable of translating images into instructions.

"Pictures"—The Imaginal System

The representational system of higher primates can be used in both a visual way and a wordlike way. We need not examine the wordlike system to document proofs of differences between ape and human, for we can find profound differences in the picture-storage system itself.

Chimpanzees have extreme difficulty with photo-object matching; certainly such a difficulty is unknown in the normal child. The chimpanzee may be five to six years old before successfully matching photos and objects (such as a real shoe with its exact photo replica, etc.), whereas an eighteen-month old child can pass this test even though it has never previously been exposed to pictures. In addition, the chimpanzee shows a peculiar form of matching on initial tries, placing any photos together (such as of an orange and a shoe), and any objects together, rather than matching orange and shoe with their respective photos. Only retarded children demonstrate some version of this tendency. They match objects, such as an orange and a shoe, but the similarity ends there. They do not put pictures together and, in certain tests, can match pictures successfully with the objects they represent.

When the ape is shown familiar territory on television—filmed scenes of its home cage or the outdoor compound in which it plays—it performs just as poorly. A film of a favorite trainer, shown hiding some food in the compound, will not enable the chimpanzee to find the food. The child, for whom it is a simple matter to match pictures and objects, has no trouble seeing the relationship between the real and videotaped representations of spaces. He can go in both directions, from TV to world and from world to TV.

Is the ape's problem specifically a difficulty in transforming two-dimensional models to three-dimensional ones? His failure with both pictures and videotapes could be described in these terms. To test this possibility, we designed a three-dimensional model, a dollhouse that was an exact replica of the ape's familiar test room. In it were four miniature containers, placed on various pieces of little

FIGURE 25: *Bert peers into a dollhouse—modeled after an actual test room. He watches the trainer bait the miniature container on the miniature chair. Bert then goes to the actual test room and finds the food. On this attempt Bert was lucky. Neither he nor any of the other juveniles were able to use the dollhouse to guide them in the test room with any consistency.* PHOTOGRAPHS BY PAUL FUSCO.

furniture arranged about the small room, matching exactly the furniture and containers in the original test room. The front could be opened (quite as could the door of the test room), allowing the ape to stick its head into the dollhouse, where he was greeted by the miniature replica of his test room. Incredibly, the ape was unable to use this three-dimensional model as a guide to the real world. Such a model was no more informative of the real world for the ape than was the film presentation.

Undaunted by these failures, we attempted, with a series of carefully graded steps, to teach the ape how to use a map. First, we arranged two identical rooms, one next to the other, arranging the furniture identically in both rooms. We showed the ape which container was baited in one of the rooms, then led him to the twin room, where he promptly located the correct container. Although we knew the chimpanzee was being shown two different (if identical) rooms, we began to wonder if the animal itself knew it was being tested in two rooms (or if it was simply confusing one room with another). To find out, we sat the animal in the hall, from which it could see both rooms simultaneously; we baited a container in one room or the other, with the animal looking on, and found that when we released the animal, it always entered the appropriate, baited room. Evidently, it knew there were two rooms.

The ape's failure with normal models, such as photos, television, and dollhouses, had forced us to the extreme of using a model that was identical to the world it represented. Finding that the ape could use an identical room as a model helped us to move toward our ultimate objective—the use of a map. With that in mind, we covered the floor of the model room with a canvas sheet, placing the original furniture upon it exactly as it had been. We next reduced the size of the canvas and the size of the furniture. Then we placed the smaller replicas of the original furniture along the perimeter of the canvas (not the edge of the room), for we were trying to use the canvas as a representation of the room next door. Much to our pleasure, these reductions and changes, while at first disruptive, did not result in any permanent changes in the animals' performances. They could use the canvas with its reduced furniture as a guide to the real room. We continued reducing the size of both canvas and furniture until the furniture became no more than a kind of abstract representation of the originals, arranged on an essentially map-sized piece of canvas. While some disturbances again occurred initially, the apes' performances recovered once more.

Could we now use the map outside of its original room? And if we changed the orientation of the map, could the ape locate itself appropriately? While two of the animals continued able to use the map as a guide when the map was moved out of the room into the hall (a very minor transfer test), all four animals failed to recognize new orientations of the map, even when the map was left in the

FIGURE 26: *Jessie watches intently as Dr. Guy Woodruff places the miniature models of the furniture along the perimeter of a canvas sheet intended as a map of the test room.* PHOTOGRAPH BY PAUL FUSCO.

original room and the angle of change was a mere 45 degrees. So, our elaborate training had really accomplished very little. Not one of the animals could be given the piece of canvas in its home cage (with a mark on one of the pieces of furniture) and then travel the few steps down the hall to the test room and find there the concealed food. Even the two animals that could use the map outside the original model room failed abysmally when shown a new map of another, though familiar, space. The training, for whatever its small success, had failed completely to instill the idea of a map.

The map reading of the child, who does not need any elaborate training, is quite different. Certainly by the age of five, he or she already has grasped the idea of a map. While the ape cannot manage to adequately use even an old map of an old place, the child can surely recognize a map of her room. Indeed, she can draw pictures of familiar places—her bedroom, school playground, and

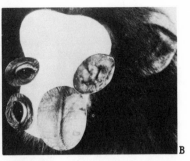

A   B

FIGURE 27: *In A, we see an example of one of Sarah's more accurate solutions to the facial puzzle. By the age of four, children can work such a puzzle, and their solutions are generally more accurate than Sarah's. Both younger children and juvenile chimpanzees have a tendency to arrange the facial pieces along the black-and-white boundary of the photograph as shown in B.*

the like—that preserve the spatial relations among key objects and thus are simple maps.

The failure in the chimpanzee's ability to recognize pictures, maps, and the like is shown further in its inability to produce drawings of simple objects. Starting with an enlarged photograph of Peony's head, we blanked out the face and gave both children and apes appropriately sized photographic pieces of Peony's eyes, nose, and mouth with which to reconstruct Peony's face. At about the age of four, children were quite successful in placing the facial features in their correct positions on the face. Of the eight chimpanzees, about half of whom had received language training, only Sarah placed the facial pieces in even approximately correct positions (she was about eight years old at the time). The other chimpanzees either stacked the pieces like blocks or arranged them in vertical or horizontal lines along the borders of the photograph. Children under the age of four also play with the pieces in the same manner; they obviously do not recognize the task as a problem in an incomplete structure.

An interesting sidelight, however, demonstrated that while Peony and Elizabeth failed, it was not because they did not recognize the specific facial features—nose, eyes, and mouth. For when

FIGURE 28: *After watching herself try on hats in a mirror* (A), *Sarah produced transformations on the face puzzle. In B, she turned both nose and mouth pieces over and placed them on the head, in a hatlike position. In C, she arranged the peel of a banana in a hatlike manner. She did not "make hats" before viewing herself wearing them. And when she no longer had the opportunity to try hats on, she ceased the "making" of hats. D is a puzzle transfor-*

they were given match-to-sample tests using a flashlight, a bell, and a flower as samples—with the various photographic features as alternatives—they surprised us by correctly matching eyes to flashlight, ear to bell, and nose to flower.

This knowledge of what function goes with what piece reminds us of the ape's success in tests requiring it to match a seed from say a pear to its appropriate stem, the stem of an apple to the appropriate fruit, and so on, all of which the ape manages very credit-

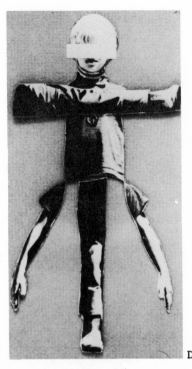

D

mation carried out by a seven-year-old child. Given arms, legs, torso, and head photos, the child first worked the puzzle correctly; then he improvised. Replacing the legs with the arms, he then placed one leg across the shoulders, the other in the position of the penis. In carrying out these transformations, both Sarah and the child make their own symbols—using one thing to stand for something else. Called "symbolic play," this is common in children but not in apes.

ably. But these tests, like the matching ones described above, do not require the reconstruction of a whole item. In order to reconstruct a whole from its parts, one must not only be able to recognize the parts but also know the relations among them. An adequate reconstruction can only be made by correctly placing the eyes a reasonable distance apart, the nose in a midline between the eyes, and so on. But the chimpanzee has difficulty making an analysis of either face or fruit that will assist it in placing one facial piece at an appropriate distance and angle from another piece. This perplexes us. For while the chimpanzee can associate all the correct parts of particular fruits, and Peony and Elizabeth could indicate the proper functions of the various parts of the face, they could not build a recognizable face even when given the parts. It is of little wonder, then, that Sarah failed to build a face from scratch when invited to do so with a wad of clay (though she, at least, consistently separated the wad into pieces). Chimpanzees do not in fact sculpt, draw, or paint representational figures or scenes. Some researchers have credited them with abstract art, but if abstract art is a less representational level of art practiced by those who have already mastered the representational, calling the ape's use of lines, colors, and dribbles abstract art is not only a dubious statement, but also a silly one.

The fashioning of representational, visual forms is no less universal in human beings than is language. The chimpanzee's incapacity is not the lack of a motor skill, not a physical inability to draw. Rather, it is the inability to analyze complex objects into their parts and to understand the relations among them. The child develops a set of primitive elements (without any training), which he combines, as he matures, into more complex elements before he approaches representational drawing. Children start with about twenty extremely simple elements, such as a dot, horizontal line, vertical line, arc, wavy line. They then combine these primitive elements into more complex forms, such as horizontal and vertical lines drawn in a cross, two arcs into a circle, two crosses drawn side by side. In all cultures, children pass through roughly four stages, always building more complex figures from simpler ones, before producing their first representational drawings. These are typically faces and suns. An examination of chimpanzee marking skills does not show any such natural development. In other words, contrary to popular belief, though we may give the chimpanzee a human hand in place of its own, it will not draw representational pictures—not any more than the chimpanzee will, if given a human larynx, produce even the most childish conversation.

"Words"—The Abstract System

Our best evidence for abstraction in chimpanzees comes from the same/different tests, and from the tests on analogical reasoning. All the language-trained animals, Sarah, Elizabeth, and Peony, received and passed the same/different tests. However, the tests on analogical reasoning were given long after Elizabeth and Peony had left the laboratory. Sarah was the only language-trained chimpanzee to receive the thorough and comprehensive set of analogies tests; she passed them very creditably.

Why is this our best evidence for abstraction? Animals are not relying on physical similarity when they make these kinds of judgments. An animal could judge two oranges similar or two elephants similar, making both judgments on the basis of physical resemblance. But to judge the relationship between the two relations—oranges in one case, elephants in the other—one cannot merely use physical appearance. There is no sense in which the relation between the two oranges looks like the relation between the two elephants. To judge this more comprehensive relation—a relation between relations—the ape must use the abstract judgment of same/different.

Even more incisive evidence for the ability to store information in a wordlike way can be seen in the analogies test. While the same/different tests simply ask the animal to make judgments about similarities, analogies require far more. In the tests we gave Sarah, the relations involved changes in size, color, shape, and marking in the figural analogies. In the functional cases, Sarah received examples of such actions as cutting, opening, marking. Sarah indicated she had an abstract idea of opening by judging the relation of can opener to can the same as that of key to lock.

And Vice Versa

The human ideas of "and" and "or" seem notoriously abstract. We encounter them only in sentences, joining one clause to another, enumerating verbal lists, and so on. Of all our words, these give the impression of being impervious to any kind of pictorial representation. Yet even here, we can show how at least part of the job done by these most abstract forms can be represented by pictures or images.

The study of formal reasoning that is based on connectives such as "and-or, if-then," in one case, and on quantifiers such as "all, some, none" in the other, has typically been relegated to the classroom; most of us have had our initial introduction to this seemingly esoteric field by way of the lesson to be learned from combining the mortality of men with the fact that Socrates is a man ("all men are mortal, Socrates is a man, therefore . . ."). All our models of reasoning have depended on clear language formulations; so it is important for us to emphasize here that reasoning is commonly used in everyday life, even outside the classroom, and does not depend at all for its existence on such formulations.

Even the chimpanzee has been shown to use reason in trying to obtain food that is out of reach. More recently, we have found that our juvenile chimpanzees could solve problems of appreciably greater difficulty. The animals were required to combine two separate pieces of information in order to reach a correct solution. That is, we devised a simple experiment that would exemplify the conditions of "and" and "or."

Representing "and/or" with Pictures

By establishing two food concessions in the chimpanzees' compound, we managed to represent the notions of "and" and "or." Sometimes both concessions contained food (the "and" condition); at other times, one concession only—not the other concession (the "or" condition). In every test, the chimpanzee was seated behind a barrier that prevented it from seeing the compound. In one test, the trainer approached the animal while carrying food in both hands. When the chimpanzee had noted this generous state of affairs, the trainer departed for the compound, returning to show the chimpanzee, a bit later, that he was now empty-handed. In another test, the trainer behaved in precisely the same way, except that he approached the animal with food in only one hand. In both cases, the ape was released into the compound shortly after seeing his empty-handed trainer.

To our pleasure, the ape responded quite differently during its release time, depending upon what it had witnessed earlier. At the end of the first test (the case of "and"), the animal went to one concession, took the food, then promptly ran over to the other concession from which it also gathered the food. At the end of the second test (the case of "or"), if the animal found food in the first concession it checked, it did not visit the other but returned to the seat behind the barrier. If, however, it did not find food in the first concession, it promptly went to the other concession. Even this difference in animal behavior seems to capture the flavor of everyday reasoning—of the difference between "and" and "or." In the next study, the animal was required to combine two pieces of information, and the ape's ability to make a distinction between the logical terms is demonstrated even more forcefully.

After seeing the trainer who had carried the two pieces of food return empty handed, the animal was taken by the trainer to one of the concessions. There it witnessed the trainer remove the food. The ape was then returned to its seat and released into the compound shortly thereafter. The animal moved promptly and certainly to the concession from which food was not removed. In the alternate case, after seeing the trainer who had carried only one piece of food return empty-handed, the animal was taken to witness the trainer remove food from the concession. Now when the animal was released, it did not go into the compound at all but loitered in the vicinity of its seat.

These two pictorial situations can be described in a clear but abstract way. In the first experimental case, food is at A and B; then food is removed from A; therefore, food is only at B. In the second case, food is at A (or B); then it is removed from A (or B); therefore, there is no food. The chimpanzee, acting on something like this kind of information, approaches *B* for food in the first test but ignores both concessions in the second test. Does the chimpanzee describe to itself the test situation in an abstract fashion? We face, once again, the same perplexing question we asked ourselves concerning the chimpanzee's management of physical models: does the animal represent a situation to itself in precisely the same visual terms as those observed? Or can it translate that information into something more abstract? If the information is left in the original, visual form, then the ape must have some system for combining the visual forms.

The animal, for example, could represent "and" with a single picture showing food at both concessions. But to represent "or" it would need two pictures, one showing food at one concession, not the other; another showing food at the "other" not the "one." While the logical statement has no preferences concerning which concession in the case of "or" contains the food, an animal might have a bias in favor of, say, the left concession over the right. If this is so, the image of food in the left concession would appear first (or the two could occur at the same time, with the preferred one larger). How an individual would use such images to carry on a process of reasoning is not known. If we assume that individuals reason in an abstract form, we are at a similar state of ignorance, for we have no idea how abstract reasoning occurs either. No matter how information is stored, whether in images or more abstractly, we need to provide a system for the use of this information. And we have no better clarification of the steps involved in reasoning by abstraction than in reasoning with the use of pictures or images.

Transfer tests could help us to determine if information is being stored with images or in a more abstract form. Merely to change the ape's location (say from compound to test room), or the containers (say from tin cans to teacups) would tell us little. For if the chimpanzee still performed correctly, he could easily be using images. We need to change not merely the setting but the very nature of the problem.

Bert can hear voices from the room next to his. From the amount of noise, he can tell if both Sadie and Luvie are in or if only one is there; but not which one. When the noise volume is low, we might say that Bert's information could be represented as either "Sadie is there," or "Luvie is there." We then arrange to show Bert that Luvie is in the hall and ask him who is in the next room. Bert's correct answer would demonstrate a kind of reasoning such that: it is either Sadie or Luvie; it is not Luvie, therefore it must be Sadie. However, this problem, like the one in the compound, could still be solved with images. But suppose that training Bert on the problem of locating food in the compound helped him solve the problem of determining who was in the next room? This certainly could not be the case if Bert were using images in solving these problems, for the images in the two cases must be very different. Practice on the one problem could be a benefit to the other only if one and the same abstraction were being used in solving both problems. We would expect this benefit with human adults, for they certainly have the concepts of "and" and "or" and would apply them in both cases. Children have these concepts, but they do not apply them in all cases. With the chimpanzees it is not yet clear whether they even have the abstract concepts, let alone can use them consistently.

Chapter Seven ■

Who
Has
Language?

■ Children as young as two-and-a-half to three years of age can be observed to make systematic deletions in their sentences. Sometimes the child will say such things as, "Daddy come home," at other times, "Come home." The child may sometimes say, "Where did Mommy go?" and at other times, "Where go?" Notice, these are observed deletions, not inferred ones. We have both forms of the sentence and can observe that a word has been omitted from one of the forms of the sentence.

Does the child delete just any word, or is there something distinctive about the words the child treats in this fashion? Are they words with particular kinds of meanings? Words that appear in a particular position in the sentence? Or words that have been in the child's lexicon for so long a time that the child feels comfortable ignoring them? It turns out that none of these hypotheses is correct; the words the child deletes are, in fact, the subject of the sentence.

To characterize the child's performance, we must, in some sense, credit the child with a knowledge of what a sentence is—for the concept of subject is defined by (or presupposes) the concept of sentence. Of course, we do not mean that the child has the kind of access to this knowledge that an adult has. No doubt the child could not sort his or her own productions into two piles of sentences; ones in which the subject was deleted and ones in which the subject remained. Nor could the child identify sentences in which he had deleted the subject of the sentence rather than the predicate or direct object. Nevertheless, the child's ability to delete the subject, on a systematic basis, requires us to credit him with some knowledge of the structure of a sentence.

The chimpanzee does not exhibit a similar performance; there is no evidence of any systematic appreciation for grammatical distinctions. While we find evidence for semantic distinctions, distinctions in the meaning of words, syntactic distinctions are not within the capacity of the chimpanzee.

The child's performance is uncommon because, while syntactic distinctions are almost always connected to semantic ones, the child's deletion strategy is not. It is based on purely syntactic considerations; the subject alone is deleted. Further, the evidence for deletion is detected at a time when the child is extremely young, and her grammatical system is weak—so weak, in fact, that her sentences consist of no more than three or four words. As a rule, in order to find evidence of syntactic knowledge, we have had to rely on the human adult who is already using a strong grammatical system. If, however, the only evidence we had for grammatical sensitivity were supplied by the human adult, we would hardly be surprised by the lack of it in the chimpanzee. After all, even the most avid proponents for language in the ape never supposed that the ape would be capable of adult grammar. But, the evidence we have makes it clear that even the brightest ape can acquire not even so much as the weak grammatical system exhibited by very young children.

The chimpanzee does, on the other hand, demonstrate that it recognizes the equivalence between the world on the one hand, and its plastic language on the other. Not only did Sarah answer correctly such questions as "Is red on green?" referring to two colored cards, she sometimes even changed the actual arrangement of the cards so that she could answer "yes" to the question. Not only did she recognize the correspondence between world and words, she preferred to see a conformity between the two. On 70 percent of trials in which world and words did not agree, Sarah compliantly answered no, but on 30 percent she altered the objective situation and then answered yes.

A similar recognition of the correspondence between the two systems, world and word, is seen when Elizabeth describes her own behavior. About 25 percent of the time, after she had carried out a simple act such as cutting an apple, she would describe her action with her plastic words as: "Elizabeth cut apple." Her descriptions of her own behavior were always correct (in contrast to her descriptions of that of others' behavior, which were about 70 percent correct). In both these cases, we have clear evidence that the chimpanzee recognizes a correspondence between the language system and actual events in the world.

Bees are said by some to have language, but the evidence for this claim does not include any examples of an ability to judge the agreement between the world and messages about the world. Instead, the bee responds automatically to the information encoded in its dance; it expresses no doubts about the accuracy of the dance, asks no questions of the dancer.

Interestingly, the ability to see a correspondence between events in the real world and plastic words is just another, albeit more powerful, example of the relation between relations. The simplest example of this capacity is that of same/different. Not until we started examining this apparently simple judgment with some care did we see that its implications are even more extensive than those we mentioned in chapter 2.

When an animal is shown two apples and two bananas and is asked to indicate the relation between them, it identifies the relation between the apples as the *same* and that between the bananas as the *same*. It then says that "same" describes the two relations.

The animal does precisely that in answering questions. It identifies the relation between a red and green card as *on*, understands the construction describing the relation between the red and green card as *on*, and then replies yes—a judgment of sameness now rendered on the relation between written and real versions of on, rather than between instances of same. Species that cannot make same/different judgments obviously cannot attain even the most basic requirements for language use.

Suppose a scout bee gathered information about the direction and distance of a food source and encoded this information into its dance. Could this same bee, shown its own subsequent dance, judge whether or not it accurately represented the direction and distance of the source of food? Could the bee, in other words, recognize the dance as a representation of its own knowledge? These are the kinds of tests we need for the bee. If a bee could judge between the real situation (the perceived events) and a representation of the situation (the produced dance), we could interrogate the bee just as we interrogate the ape. And species that can be interrogated are well on their way toward making judgments about the accuracy of representational statements, whether they be dances, plastic symbols, or spoken words.

While it would be desirable to perform tests of this kind on the bee, we already have substantial grounds on which to anticipate their results. The judgments in question are those of same/different and, as we know, presuppose an abstract representational system. In addition, making judgments about the reliability of a speaker presupposes attributing intention to him. These are both extremely high-order competences; evidence for having them requires passing some extraordinarily demanding tests, such as tests for same/different, which we have discussed above. In addition, recall the video tests given Sarah to determine the attribution of intention. The bee would have to pass the same kind of tests in order to be interrogated as to its judgment of representational statements.

There is a mistaken tendency to equate word order with syntax; to suppose that, because the animal can distinguish, say, "red on green" from "green on red," or even "Mary give Sarah apple" from

"Sarah give Mary apple," it must know what a sentence is. But word order is not the equivalent of syntax.

A common informal definition of sentence such as "a thought expressed in words" describes what the chimpanzee can do—for chimpanzees have both thoughts and words. They even can learn to observe word order, distinguishing "Mary give Sarah ice cream" from "Sarah give Mary ice cream." But a sentence is not merely a thought expressed in words; it is a thought expressed in words arranged in a sequence chosen from indefinitely many other possible sequences. Provided you have genuine knowledge of a sentence, you can express the thought in a countless variety of ways. Even so simple a thought as "dinner time" can be announced to a child as "eat-eat time," to a husband as "Time for dinner," to guests as "Dinner is now being served," to Rover as "Come and get it," and so on.

Notice that we have carefully referred to the chimpanzee's strings of plastic words as constructions, not sentences. A sentence has an internal organization—its parts are defined relative to one another and independent of the world. The construction does not have an internal organization; its parts are not defined relative to one another; it borrows its organization from the world itself.

For example, in the giving exchange, we mapped a word-order rule in which the donor was mentioned first, the action next with the object following, and the recipient last; that is, "Mary give apple Sarah." In another construction, "red on green," we adopted a rule that mentioned first the object that was on the top of the other one. In yet another example, "apple beside orange," we adopted the rule of mentioning first the object that appeared to the left of the speaker.

These rules have nothing to do with one another. Obviously, a donor is not a top card, nor is she an object that appears to the left of the speaker. The construction rule is determined entirely by the various physical situations; the constructions themselves have no internal organization. A human looking at these three constructions would dissent heartily with our conclusions, insisting that they do have much in common because "Mary," "red," and "orange" are all subjects of their respective sentences.

Of course, we agree, they are. But is Sarah equipped with this knowledge? If Sarah could, after being trained on one sentence more easily learn another sentence (transfer across cases), we would have to credit her with knowledge of "subject" of a sentence. To credit the ape with knowledge of the concepts "and" and "or," we needed a transfer across the case of "food in both containers or in one only" to the case of "both chimpanzees in the cage next door or only one." While these seem to be incomparable abstractions, "and" and "or" pale when compared to the abstractions of subject and predicate.

Memory and the Representation of Words

Of all the prerequisites for language, none is more vital, though more easily overlooked, than memory; yet language is possible only because of memory. For it is not the objects, the actions, or the properties that we perceive in the real world that are associated with words, but their representation in memory that makes language possible.

We do not know whether species store information differently, that is, whether the quality or power of memory varies with each species. If so, there is a limit on the effectiveness of the language they can acquire. For the name of an item is only as informative as that of the representation that is stored and associated with it. An animal that could not retain representations in memory of the world it sees and has seen would not have the capacity of language.

We commonly speak of the "power of the word." No doubt the phrase celebrates how extraordinarily well a word can be used to substitute for a referent. This substitution is successful because we can use the words as a device to retrieve information; the word has power because the name can provide us as much information as the actual presence of the object. People who know a great deal about, say, giraffes, can sit and talk about them informatively even if giraffes are not loitering in the immediate vicinity.

What is the difference between a judgment that an animal makes about objects it simply perceives, and the judgment it makes about an object it must reconstruct from memory? Is the one, such as seeing an apple, more reliable than remembering an apple? When an animal is asked to match an object with a patch of color—say, an apple and a red card—both the object and the color visibly match. The animal, in answering correctly, demonstrates the acuity of its perception. If, however, we leave the card red but paint the apple white, a perceptual match no longer exists. Only if the animal can reconstruct an apple from memory can it decide, in this case, whether the red card is, or is not, the color of the apple.

To test our chimpanzees on their knowledge of the fruit they were fed every day, we divided eight fruits into four components— wedge, stem, peel, and seed—and two features—color and shape. Taste was added as the one nonvisual feature. We assumed that if the animal knew a great deal about an object, it would need only a small sample to identify it. If given only a stem or a seed (or even a taste), for instance, an animal with a strong representational system could reconstruct the complete fruit when shown only a sample of it. A rich memory system would provide more information than a poor one. Therefore, it was important to know how much information apes could store about their world of fruit.

Our four language-trained chimpanzees were each given about two dozen tests on varieties of fruits. On each test, they received either the whole fruit or one of its seven parts as the sample, and two other parts as alternatives. For example, the seed of an apple might be the sample, with a red patch and a yellow patch the alternatives; or the animal might be given a taste of banana and then offered as alternatives two white cardboard forms—one the shape of a banana, the other that of a cherry. If the animal both recognized an apple seed and knew the color of an apple, it could choose correctly; as it could if it recognized the taste of banana and knew its shape. Sarah's results were especially impressive; she could use every cue correctly. The other three animals, though generally accurate, were more successful with some parts than with others. As for them, it was possible to rank-order the informativeness of the several cues. Not surprisingly, the whole fruit was most informative.

But color and peel were almost as effective; the chimpanzees were almost as able to match the seed, stem, shape, and wedge to a patch of color as to the whole fruit itself. Taste was less informative, followed by a tie between shape, wedge, and stem. Trailing all the components was the seed, the least informative cue of all. Differences of this kind do not change the main finding. Chimpanzees, Sarah especially, but the other animals as well, can store detailed representations of items such as fruit.

In the next test series, in order to test the power of the chimpanzee's words, actual parts of the various fruit were shown as samples. This time, the names of the fruit were given as the alternatives. Perhaps the seed of an apple, with the names "apple" and "cherry." The match-to-sample test results were clear-cut: words were fully as informative for the animal as were the actual fruit. Even more surprising, the words for fruit were far more informative for the chimpanzees than were actual fruit segments, even though most of their meals consisted of segments of fruit. That is, when given names of the fruit, they could more successfully identify parts of the fruit than when given actual parts of the fruit. In general, the name was about as informative as the whole fruit itself. In the ape, it seems the word substitutes vigorously for its referent.

When asked "What is the color of apple" or "What is the shape of cracker," Sarah can provide the correct answers "red" and "square," even though neither the apple nor the cracker is present. The results of the word-memory test explain why such an outcome is possible. The ape is capable of storing a good representation of the fruit and can retrieve the stored representation with the plastic word.

At an early stage of language training, Sarah used words to request preferred items that were not present. Presumably, she could do so because seeing the plastic word for apple enabled her to "picture" an apple. Similarly, when given the word for chocolate in teaching her the word for brown (as in "brown color of chocolate"), Sarah could use her word for chocolate as an aid in picturing chocolate. So that when told to "take brown," she correctly took the brown rather than some other colored disk. On the other hand, while Sarah requested one fruit or another when the actual fruits were nowhere present, she never did, say, request a short visit from the absent Bert.

Compare the information retrieving power of the word in the chimpanzee with that of a barely two-year-old child of our close acquaintance. The child's grandmother is in the kitchen preparing a sandwich for his lunch as he stands close by, observing her from a kitchen chair. He points to a jar of Hellmann's Real Mayonnaise she is using and says, "Mammy have some dat home." In doing so, he refers to his mother, who is out shopping, to his home, which is about a mile away, and to a jar that recalls his mother's jar of mayonnaise. On another occasion, he looks at a painting and remarks, "Danny have one Danny room"; again, when he is given the use of a small hammer, he volunteers the information that "Daddy use big one, Danny use little one."

All these sentences are produced spontaneously. The child does not merely refer to an object but to facts concerning the object, such as where it is to be found (home), to whom it belongs (Danny), who uses it (Daddy), and its relative size (small and big).

The chimpanzee is able to use words to retrieve its knowledge about the real objects that words represent, as when it describes, for instance, a blue plastic triangle, which is its word for apple, as being red and round, with a stem. It can recognize such statements as "red on green" as representing a red card on a green one. The bee cannot learn, or retrieve information from, words. The bee has never been shown to recognize its dance (which is said to contain information concerning the location of food) as a representation of its own information. To be credited with language, the bee must be able to recognize its dance sequence as representing actual events, quite as the chimpanzee recognizes its word sequence "red on green" as doing so.

Language is possible only because of memory, since it is not the actual objects, actions, or properties per se that are associated with words but their representation in memory that makes language feasible. An animal must retain representation of its visual world in memory in order to be capable of language. The quality and power of stored information varies. The level of the language a creature can acquire is, in part, NO MORE INFORMATIVE THAN THE REPRESENTATION that is stored and associated with language.

While the linguist on the one hand, and the semiotician on the other, claim priority for syntax and intention as basic to language, we emphasize the crucial role of representational capacity. We view representational capacity as the ability to judge the relationship between actual events and representations of them. If language is viewed as a family of representational systems, we rescue it from ethnocentric limitations and include the language of apes within the family.

Human competence is complex and frequently elusive. Certainly one of the easiest ways to demonstrate that a chimpanzee can do what a human can is to start with a simplified account of human beings. We do not feel we have fallen into such a trap. While we have shown certain capacities for language that apes share with us, we have also emphasized the limitations found in the representational system of apes. The sentence is the most abstract representation of which humans are capable and, as such, is far beyond the capacity of the chimpanzee.

Chapter Eight ■

Characteristics of an Upgraded Mind

■ We tend to think of animals as judging objects on the basis of appearance—not on the basis of meaning—because we consider animals to store "pictures" and to compare the pictures with what they see. We expect animals to recognize the similarity between physically comparable objects. If an animal learned to peck, say, a yellow grain, we would not be surprised to find it pecking at another, similar grain. We would be surprised, however, if the animal pecked at pictures of a grandfather and a grandchild, a sheep and a lamb, a frog and a tadpole—but did not peck at pictures of a bull and a cow, two lambs, and two grandfathers. We do not credit animals with a knowledge of the relation old-young or mature-immature. Abstract judgments, we believe, are within the human realm only, as is standing upright.

The Evidence for Abstraction

The curious consequence of language training seems to be that it weakens the characteristic difference between person and animal. For it appears to convert an animal with a strong bias for responding to appearances into one that can respond on an abstract basis. The proportions test that we described in chapter 2 provides a straightforward example of this distinction. Sarah and the four juveniles were given match-to-sample tests involving varying proportions of fruit (in one case) and glass cylinders filled with water (in another). All the animals could judge correctly when shown, say, a quarter of an apple as the sample, and one-quarter and three-quarters of an apple as alternatives; they did equally well when given, say, a half-filled glass cylinder as the sample, and a half- and a quarter-filled cylinder as alternatives. These are judgments that can be made in terms of sheer appearance. Two quartered apples look alike, as do two half-filled glass cylinders. But in the next step of the experiment, when the animals are given, say, a three-quarters-filled glass cylinder as a sample and three-quarters and one-half of an apple as alternatives, the correct alternative could no longer be selected on the basis of appearance. The equivalence of the three-quartered apple and the three-quarters-filled cylinder had to be arrived at on some other grounds. We suggested the analogy as one strategy (perhaps the only one) for an animal that does not have names for the proportions. On that theory, the ape would use a visual version of the following ratio: three-quarters of an apple is to a whole apple as a three-quarters-filled glass is to a completely full glass. Sarah passed this test, but all four juveniles failed it. Language training probably made the difference.

Sarah could make not only same/different judgments using her plastic words, but also an equivalent of such a judgment without the use of words. In these tests, she was given pairs of items in a match-to-sample format. For instance, rather than using orange as the sample, with banana and orange as alternatives, we used a pair of oranges as the sample, a pair of apples as the correct alternative, and a banana and an apple as the incorrect alternative. Conversely, we used orange and apple as the sample, pineapple and pear as the correct alternative, and two pears as the incorrect one. Sarah passed this test with flying colors, sixty-one correct in sixty-four trials on her very first series. No plastic words were used in these tests, but it is quite clear that the abstract judgment of same in the first case, and different in the second, must somehow be used by the ape to choose the correct matches.

The four non-language-trained juveniles not only failed these tests, they continue to fail them in spite of repeated drill and reward. These animals can, however, do what is called generalized match-to-sample, which looks like the following for our chimpanzees. In the laboratory, the animals were taught traditional matching on the basis of assorted toys. That is, they were given a toy car as sample, a toy car and a doll as alternatives; after only a few examples of this kind, they transferred their skill at matching to a variety of new toys. The successful transfer is referred to as generalized matching. This generalized matching is something the pigeon can also do, though less well.

Did the apes' success on matching toys mean they could match other kinds of items? To find out, we took both toys and chimpanzees into the compound. Sometimes we used plants (growing in the compound) as samples; sometimes the toys. We used a toy and a plant as alternatives in both kinds of trials. To our dismay, the juveniles did not match under these conditions. The animals chose the toy on every trial—even though they were perfectly familiar with these plants, having eaten them every summer for about five years. In matching the toys, they had learned not only to match but also to classify—the toy as a particular kind of thing—and to take things that belonged to the class. Their failure in the compound, however, was transient. We eliminated the toys, giving them a few trials on which plants alone were both sample and alternatives. They succeeded immediately, matching a dandelion with a dandelion, and so on. When next we restored the toys and tried the original compound test—toy as sample on some trials, plant on

FIGURE 29: *This is a match-to-sample test in the compound. In A, Dr. Guy Woodruff shows a juvenile the sample of a plant. In B, the animal searches the compound grounds for a match. The animal places the match in the bucket in C; in D, it is rewarded.* PHOTOGRAPHS BY PAUL FUSCO.

others—they matched correctly from the first trial, plant with plant, toy with toy.

Will the juveniles continue to respond to new items as they did when plants were first introduced—when they chose old items (toys) on the basis of their appearance, rather than new items (plants) on the basis of their sameness? How many more new classes of items would the juveniles need before responding to sameness from the beginning? Suppose we had introduced clothing: for example, shoes, boots, hats, sweaters. When a shoe was the sample, would they match a shoe to it or pick the toy or plant? And would we again need to eliminate the plants (and toys) as an intermediate step, giving them only clothes to match before returning them to problems that included both old and new classes?

The same/different judgments that Sarah made without the use of plastic words are the equivalent of analogies. For example, "apple/apple is to orange/orange" is already an analogy. Though one restricted to sameness (analogies can involve not only sameness but a variety of relations), it is still one of the simplest kinds of abstractions—the litmus test we can use with animals without words. Sarah, of course, went on to pass analogies tests that involved not only complex images but functional relations as well. The non-language-trained animals could not pass even the simplest kind of analogy. Sarah, Peony, and Elizabeth passed the action tests right from the start. They recognized knives, bowls of liquid, and pencils as instruments that changed an object from its initial to its final state. They also recognized that a cut apple, wet sponge, and marked paper were the "final states" to which apple, sponge, and paper were brought by the instruments in question. They recognized the relations of cutting, wetting, and marking in the case of not only familiar items but also others that were unfamiliar and, indeed, anomalous. For example, they chose a knife as the instrument that was used to sever ping-pong balls, a pencil as the instrument that was used to write on apples. In contrast, the four juveniles failed the same tests—and continued to fail consistently over four test replications.

Sarah not only recognized the instrument that brought about a particular change, but distinguished one aspect of a change from another, recognizing that a change in shape did not change quantity. In other words, Sarah conserved both liquid and solid quantity; but the juveniles (tested with a match-to-sample procedure to accommodate their lack of words) did not. Although the ability to conserve is not fully understood, tradition attributes that ability to conceptual-level responding.

The young child, who does not conserve, appears to believe that, for example, clay that is compressed in shape is also reduced in amount, basing the judgment on appearance. While the older child does not make this error, evidently discounting appearance and basing his judgment on what he "knows" and infers. It seems that Sarah passed her tests of conservation with these considerations in her "mind"; the juveniles failed the tests.

Neither Age nor Intelligence . . .

There was a considerable age difference between Sarah and the juveniles when they were given these tests. Sarah was about fourteen years old, the juveniles only five or six. Fortunately, we have other tests where the language- and non-language-trained animals were the same age when tested. Peony and Elizabeth, on the one hand, and the four juveniles, on the other, were given the action tests when about five years of age. Similarly, these two groups were the same age when the former passed the same/different tests, and the latter failed the nonlanguage version of same/different judgments.

But perhaps neither maturity nor language training is responsible for these differences. Perhaps the non-language-trained animals were simply dull. Chimpanzees do, indeed, differ in intelligence; in fact, they differ widely. But the four juveniles did not compare unfavorably with the language-trained animals in this regard. Although Sarah is a bright animal, so is Jessie (one of the juveniles). Jessie is, in fact, more full of "wit" than her peers.

She is, for instance, our only chimpanzee to demonstrate unsolicited giving. She fills her mouth from the drinking fountain located at the edge of the compound and crosses the compound to a high beam, where her friend Luvie is basking in the sun. Jessie ascends to the reclining Luvie and delivers the drink directly into Luvie's mouth. Jessie was also the only one of the juveniles to remove the blindfold from her trainer's face (without hesitation, on the first trial) in order to lead him speedily to a locked box filled with food. The trainer, of course, had the only key to the lock. Moreover, she did not remove the "blindfold" when it was placed not over the trainer's eyes but around his mouth or hair. The other juveniles dragged the blindfolded trainer arduously across the compound at every trial, never even attempting to remove his blindfold.

Nevertheless, on tests that required an abstract level of responding, Jessie failed just as resoundingly as did her confreres. Finally, there is the other side of the coin to consider: that is, Peony—not

FIGURE 30: *Bert confronts the blindfolded trainer and does not remove the blindfold. He grasps the key chain around the trainer's neck and laboriously leads the trainer to the locked cache on top of the wooden platform.*

exactly a stellar pupil. She passed the same action tests that Jessie failed. Neither brightness nor age seems to account for the particular difference between the two groups.

. . . Nor Reasoning

Granted that it is language training, and not age or intelligence, that makes the difference, can we be sure of what the difference is? Does the distinction between image and abstraction—between the appearance of something and its meaning—really capture the difference between animals with and without language training? Perhaps what is at stake is reasoning rather than representation. Could language training affect the ability to reason, thus enhancing the ape's capacity to pass the tests?

We considered this a serious possibility and tested the non-

FIGURE 31: *Jessie almost immediately removes the trainer's blindfold, and he follows her easily.* PHOTOGRAPHS THESE TWO PAGES BY PAUL FUSCO.

language-trained animals on two kinds of reasoning "and" versus "or" and transitive inference. Transitive inference is a form of reasoning in which an individual is first told (or shown), for example, that A is bigger than B; B is bigger than C. The individual is then asked about the relation between A and C. If he answers that A is bigger than C (not only in this particular case, of course, but in others like it), he is credited with the ability to make transitive inference. For, although he has no direct knowledge about a relation between A and C, he can infer the relation because of what he knows of A and B, and of B and C.

All the non-language-trained animals passed the "and/or" tests, and Sadie, the oldest of the group, passed the transitive inference test. Although the concepts of "and" and "or" are indeed abstractions, we saw that comparable distinctions could be made on the basis of images. To represent "and," the ape needs an image of food in both concessions in the compound; to represent "or," an image of food in one concession only, the other concession being empty. Representations of this kind could account nicely for the ape's solution of these tests. In the one case, when released into the compound, the animal did not go to the second concession if it found food in the first; while in the other, the animal went to both concessions.

How individuals make transitive inferences, precisely, is still open to conjecture. There was a time when it was believed that children in a certain age range could not make such inferences; but then it turned out that they had trouble remembering what they had been told about A and B, B and C. Once they were well drilled on the initial relations, they managed the test nicely. The problem was not one of "poor" reasoning, but poor memory. It was also once considered that language or abstract representation was essential for transitive inference; but since such inference can be made by non-language-trained apes, and even perhaps monkeys, this cannot be true. Moreover, although humans do use language in making transitive inferences, they do not use it on all occasions. People also use images sometimes, in which A is visualized as being larger than B, B as larger than C—from which they can "see" that A is larger than C. Language training does not confer the ability to reason, for both kinds of reasoning can be done with images. The non-language-trained apes can do both "and" versus "or", and transitive inference.

... But Language Training Itself

Exactly how does language training affect the representational system? The training is a two-edged sword, suppressing forms of responding that would ordinarily interfere with abstract-level responding and, at the same time, enhancing the use of abstraction. The major interfering form is that of physical matching. Grouping objects that are physically alike is a primitive form of responding in the primate (both in early development and in vigor). A simple demonstration of this disposition can be clearly seen in the following situation. Highlight one object, such as an apple, and below the apple, place other objects, such as another apple, a red patch, a banana, and a shoe. Permit the individual to choose from all the items and simply record the order of his choice.

In whatever order you present the objects, both child and chimpanzee will choose the objects in this particular order: apple (identity), red patch (feature), banana (category), shoe (irrelevant). As the child grows older, features (color, as an example) will replace identity as the first choice. This is true in the ape as well, although the age at which identity loses its strength is far later in the ape. Language training much weakens the potent disposition to do identity matching (apple to apple), because the word that must replace the interrogative particle in the chimpanzee's "questions" is not a word that matches some element in the string.

Data from our action tests show some support for the claims we are making. Of the incorrect choices made by the four non-language-trained animals, approximately 30 percent to 40 percent were clearly cases of physical matching. For instance, in the test sequence "apple, blank, cut apple" though knife was the correct choice, pencil was most often chosen. Of the alternatives offered, only the pencil was red.

However, the success of language training cannot be based entirely on its value as a suppressor of certain kinds of competing natural tendencies. Even animals that fail the action tests (by their persistence in matching items that are physically alike) will continue to respond unreliably when every effort has been made to eliminate any possible "matching" choices in the alternatives. Although it would be simplest to consider that language training plays only a suppressive role, the data do not support this simplistic view.

How Language Training Enhances Abstraction

How does teaching language to the ape enhance its use of abstraction? To answer this question, we must clarify the differences between the two kinds of representational systems—those using pictures and those using words. And we must show why there are two kinds and why we grant both to exceedingly few species—humans, apes, and monkeys—and only one to all other species.

Of the many items to be found in the world, nearly all of them can be represented by pictures. To begin with, everything we consider to be an individual thing can surely be represented in this manner. The individual words on this page, each piece of furniture in this room, each member of your family, and so on. Parts of individual things can also be represented pictorially, for example, by placing a circle around the lampshade, around the letter *e* in the preceding word, around your son's knee, and so on. Properties, too, could be represented by a comparable system: if the picture of the red apple is meant to represent not apple but color, then mark the picture with one check; shape, two checks; size, three checks; and so on.

Is it only particular things (their parts and properties) that could be represented in this way? No. Certain groups of things could also be represented by pictures or images. For example, not one lamp but three, not one person but several, not one word but many. Do we need an individual picture of each lamp, word, or person? No, again. We could economize, using one picture to stand for the lamps, one for the words, one for the persons, because all lamps have certain features in common (as do words in English and all persons). Exactly which picture we use to stand for the group is not important. For persons, we could use either an actual picture or a stick figure. We must, however, mark the picture in a distinctive way, so that the intent of the picture is clear. For example: place a star on those pictures that represent not themselves, but the groups of which they are members.

The human infant is able to recognize groups—at six weeks of age. The child smiles at faces—not at one particular face, but at facelike configurations. A father may have thought the smile was reserved for him, only to find the infant smiling warmly at a cartoon face painted on the headboard of the crib.

If we first train the pigeon to peck at certain pictures that contain trees, and to not peck at those that do not, it will respond appropriately when shown a new set of pictures. The bird can discriminate not only trees, but humans and bodies of water. It would be as much a mistake to credit the pigeon with abstract concepts of human, tree, and water as to credit the infant with the abstract concept of face. Both are responding to patterns.

All living creatures are either born with, or acquire after some period of time, a sensitivity to certain patterns; the infant to faces, the pigeon to trees, humans, and bodies of water.

Although the infant smiles at all facelike configurations, we could train her to discriminate among the faces, rewarding her for smiling at some but not others. The infant will likely learn to discriminate among faces more quickly than among other figures, such as squares, triangles, and so forth. Pigeons are able to discriminate certain natural objects, but not among such artifacts as tables, umbrellas, and the like.

Classes (or categories) and relations are unlike groups, since they cannot be represented by pictures. Could we have an image of the class (or category) vegetable? We can easily draw pictures of lettuce, radish, or onion—of the individual vegetables, but not of the class *vegetable*. Similarly, we can draw pictures of individual toys, but not of the category *toy*.

Could we have an image of the relation "on"? It is the relation that holds between a cup and saucer, a horse and a road, a baby and her high chair. We can draw a picture (or form an image) of every specific example of "on" but cannot draw a picture of what the individual pictures, cup and saucer, horse and road, and so forth, have in common. That condition, "on" itself, makes it possible for us to classify the relation between horse and road the same as between cup and saucer, as between baby and high chair. A condition, common to a variety of individual examples that cannot itself be represented by a picture, is represented in the abstract code.

The human infant is no more likely to have abstract representation than is the pigeon—though as the infant matures, she will attain an abstract system quite naturally, without special training. The chimpanzee has a potential for abstract representation but does not attain that potential unless given special exposure, language training. The pigeon has neither abstract representation nor the potential for it—and no special experience will bestow abstract representation on the pigeon.

It does no good to ask, What does an abstraction look like? For, once we have said that it is not a picture, we have said that it cannot be characterized in terms of appearance. Abstractions are best characterized in terms of what they can do. Suppose an individual understands the relations of *on, among, if-then, cut,* and so on. Then he will be capable of at least recognizing, and perhaps producing, countless examples of each. A tree on the lawn does not look like a horse on the road, yet the individual who has an abstract representation of "on," will recognize both examples as cases of that relation.

We do not credit rats, pigeons, and nonprimates generally with abstract representation, for they give no evidence of being able to represent relations, and of being able to judge the relations between them. The simplest test for abstract representation is the nonverbal analogy (AA matches BB, but not BC), and rats and pigeons will fail such tests. Of course, rats and pigeons are affected by relations. If we feed them for responding to a key, that relationship will certainly affect them. But they have no understanding of the relationships that affect them. The relation between the response and food is one of consistent temporal order—pressing the key precedes the food, but eating the food does not precede the key (which is always there). To say that the animal understands this relation, we must show that it can distinguish between pairs of events that do and do not have this relation. It must be able to put pairs of events in which one event consistently precedes the other in one pile, and pairs in which this is not so in another pile. We know that rats and pigeons cannot do this. While all species are affected by the relations among things, exceedingly few species have abstract representations of the relations that affect them.

The language training given Sarah, Peony, and Elizabeth would be expected to enhance their ability or propensity to use abstractions, for the training was nothing so much as a specialization in the

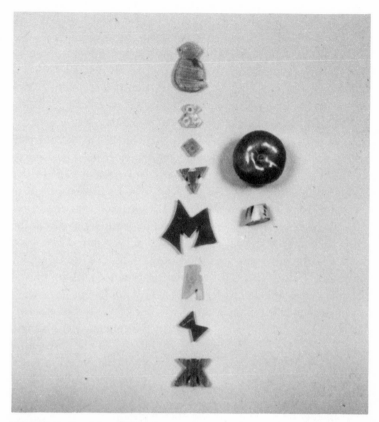

FIGURE 32: *Sarah's conditional instruction: "Sarah take banana if-then Mary no give chocolate." The relation in this example is "if-then," and the arguments are the two simple constructions "Sarah take banana" and "Mary no give chocolate." In this instruction, Sarah would be wiser to take the apple than the banana for she would therefore be rewarded with chocolate.*

identifying and naming of relations. The animal was always given a string of elements representing simple propositions, such as "apple is red," "blue is on yellow," "round shape of apple," and so on. The strings increased in complexity over the course of training to "apple is red—same—red color of apple"; "red on yellow—if-then—Sarah take chocolate"; "Sarah take apple in red dish, banana in blue dish." However, whether the strings were complex or simple, they all represented at least two-term relations.

This fact is transparent in the simple cases: *color of* is the relation; its two arguments are "red" and "apple." It is equally obvious in the longer strings, such as in the relation *if-then*, with its two arguments "red on yellow" and "Sarah take chocolate" (both arguments expressed in simple constructions).

In all tests, the animal was taught language-like skills by a procedure similar to that of sentence completion. With a few exceptions, the training-word strings always had one incomplete slot (indicated by an interrogative particle), while the other words in the string were known to the animal from prior experience. So, when the new word taught was the relation, both arguments in the string were already known. That is, "color of" was introduced when the animal already knew "red" and "apple." So that "red ? apple" asked, essentially, What is the relation between red and apple? And the animal was able to answer "red *color of* apple."

Conversely, when the new word was one of the arguments, say, "brown," both the other argument and the relation were known. That is, the word for brown was introduced in the instruction "brown color of chocolate" at a time when the animal already knew the terms for color of and chocolate. (Incidentally, this latter case also demonstrates how effectively an animal can acquire a new word merely by reading it in the construction.)

The animal was trained to perform the following operations on a wide variety of incomplete constructions: first, to remove the interrogative particle; second, to replace it with the correct word that completed the construction. This procedure constitutes an almost ideal method for interrogation.

The incomplete construction represents a question—in this definition, a question is an incomplete construction whose complete form is recognizable. It is a construction whose complete form is recognized by the ape as requiring a correct word choice.

Chimpanzees in the wild do not spend four hours a day (as Sarah did for a period of eighteen months) answering questions about the shape of crackers (? shape of cracker), the color of apples (? color of apple), proportion (? cookies are round), and the like. It is certainly surprising to find that the ape can be interrogated in this way, of course. There is very little in the natural behavior of the chimpanzee to suggest these talents are latent. Yet, when the ape is interrogated for approximately a thousand hours, can we not expect some major improvements in its subsequent mentation?

Though a potential for wordlike or abstract representation may be present in the ape, the capacity may not be well developed; its realization may depend upon special experience of a kind not found in the wild. But why should a species acquire capacities that (so far as we know) do not come to fruition in the wild and do not serve the animal after it is returned to the wild with the developed capacity? We can only reply that we know virtually nothing about the resting potential of species, that is, in what degree a species' capacities are actually realized in its natural environment.

Probably, a species does not dwell at the peak of its capacity; if it did, even slightly adverse changes in the environment could be disastrous. For example, if each leap of a brachiating species required even close to maximum effort, minor adverse changes in interbranch interval, density of predators, density of food, all would seem to doom the species. Such a species must have a reserve in strength or energy. By analogy, species that have specialized in intelligence (such as ourselves and other primates) may have reserves of this commodity. A possible test of this proposal can be found in each of these two comparisons: the one case, human groups who have and have not been to school; the other, those who are and are not literate. Both forms of experience appear to have an effect. Though they differ. Schooling appears to have the broader effect, leading the individual to recognize parallels in problems, permitting him to transfer solutions from one problem to another. Literacy, however, has a narrower effect, contributing specifically to the skills of reading only.

Neither of these changes approach the dramatic change in mentation that appears to occur when the ape is called upon to answer questions about the color of objects, their shape and size, the sameness or difference of items ranging from simple objects to pairs of short constructions, the accuracy of descriptions such as "Is red on green?" (? red on green), "Did Mary give Donna apple?" (? Mary give apple Donna), and the like. Perhaps only capacities that are in a nascent or transitional stage can be affected by excess stimulation; normal capacities may be realized by stimulation in the normal environment alone.

To produce dramatic changes, a species of one evolutionary level must act on one of another level. It may not be sufficient for humans to merely supplement the experience of other humans. Those in technologically advanced societies who bring literacy to the illiterate—schooling to the unschooled—may simply advance technologically primitive groups to what can be called base level. This may not involve any major change. In contrast, the language that the ape learned did not restore it to base level but advanced it to a new plateau. In order to achieve a comparable enhancement of human intelligence, we may require the intervention of a species whose specialization in intelligence has gone far beyond our own.

Conclusion

■ What are some of the basic questions we must ask in order to understand who we are, or, more precisely, the kind of mind we are? Essentially, we must know how we conceive of the physical world of objects and actions, and the social world of people and animals; and we must describe some of the systems that we use to represent *both of these worlds mentally*. A meaningful comparison between humans and other species can be attempted when we have similar information about the conceptions of other creatures: birds, bees, chimpanzees.

Only with genetic guidelines do we emerge with the ability to causally analyze the world, make social attributions, and store both kinds of knowledge with representations that are both imaginal and abstract. But the geneticist cannot yet tell us in what chromosomes (or genes) reside the capacity for making a causal inference, constructing a sentence, understanding an analogy, reading a map, or producing a representational drawing. But suppose he could tell us, would we then understand the nature of social attribution, causal analysis, mental representation? We cannot look to either the geneticist or the evolutionist for assistance in answering these profound and provocative questions. We must rely, instead, on the field of psychology because only psychologists raise these questions and, by experimental testing, can provide appropriate answers.

Physical world—action

Of the many events that occur around us, we see and connect only those that appear to have a cause-effect relation. If we happen to see the end of an action—for instance, an object being moved (a car entering an intersection)—we immediately infer a cause (the light has changed to green). Or, if we see the beginning of an action—for instance, an individual in motion (a painter swinging a brush back and forth)—we immediately infer an effect (a swatch of freshly painted fence).

Sarah was never taught to analyze any of the actions that were carried out every day in her immediate surroundings. Yet when shown an incomplete action—its end but not its beginning (or vice versa)—she completed the action correctly. The other language-trained chimpanzees did equally well. Children by about two-and-a half years of age can already manage these tests easily. In making correct choices, the children and animals demonstrate how well they understand the basic idea of action—recognizing the many different forms by which action can be expressed.

Most species, even probably monkeys, would look at the test, apple-blank-cut apple, and make no sense of it. At most, they would choose the alternative that physically resembled a test item, just as the non-language-trained chimpanzees chose the red pencil to go with the red apple. They would not see the idea of action hiding in the test material.

Conservation

Changes in shape, color, temperature, number of pieces, and so on, do *not* change the amount of an object. We know, for example, that tearing a slice of bread into pieces will not increase (or decrease) the amount of bread available (any more than squashing a piece of cake will reduce its quantity). Human adults know this; Sarah knows this; but children below a certain age do not.

When shown a glass of water that is poured into a narrow, tall glass (or a wide, short one), children below the age of five or six say the tall glass has more water than the short one (children seem to associate *amount* with the level of water in the glass). The non-language-trained chimpanzees, when tested similarly, made similar errors; but Sarah did not. When given her plastic words "same" and "different" and asked to judge levels and amounts of liquid, Sarah was not misled by the changed water levels resulting from the varying shapes of the glass containers. Nor was she misled by changes in the shape of clay (produced by either stretching or compressing the clay).

Unlike children under the age of about six, Sarah conserved both liquid and solid quantity. Though given no direct or special training on either the action or the conservation tests, she passed both naturally, as older children do. However, in the typical Piagetian test of conservation, the child is asked to judge number as well as liquid and solid amount. It was not possible to test Sarah on conservation of number, since she failed to make reliable same/different judgments about the number in two sets of buttons. The younger child can judge sets of numbers as *same* or *different*, then conclude that changing the space occupied by a line of buttons does not change its number. While number is salient for children at a tender age, liquid and solid amount seem not to be. Number, on the other hand, (not amount of liquid and solid) seems to lack salience for the chimpanzee.

Social World

From the human point of view, to be intelligent in the physical world is to accurately figure out "what causes what"—in the social world, to accurately determine "who intends what to whom." Because almost all human acts make the transition from elicited to intentional, it is no surprise that we interpret social behavior in terms of actions that are or are not intentional. Once we decide that an act is intentional, we try to determine the nature of the intention. When we are able to analyze the intentions of another, we feel that we truly understand the other person.

The mother chimpanzee takes her infant's arm and, extending it fully, grooms it in several places. In the nursery, the infant chimpanzee, lying in its crib, calls when some event impinges on it—a pinprick, the sight of a stranger, being separated from a crib mate—and only under these circumstances. Later, the infant whose arm has been extended for grooming will be able to extend its own arm when requesting to be groomed. But the same infant, wishing to be groomed, will not be able to *call* its absent mother when it wants to be groomed. The chimpanzee differs from the human in that neither its facial expressions nor its vocalizations become voluntary, that is, will ever be *used* to express intention.

The precocious appearance of *lying* reflects the child's ability to act intentionally (as well as to attribute intentions to others). When given truthful and deceitful information, the child can learn both to tell the difference between the two and to discount the lies. Only when given special experience do chimpanzees show any signs of being able to distinguish between their honest and deceitful trainers.

All four juvenile chimpanzees learned to suppress information when in the presence of the "selfish" trainer, who wore a bandit's mask and kept the food for himself. Two of the chimpanzees went beyond simple suppression; they actually "lied," that is, gave false information to the selfish trainer. Two animals recognized and ignored false information given them, and one animal was able to "recognize," and "tell" lies, at the appropriate times. It "lied" to the selfish trainer; recognized when that trainer was lying; then told the truth to, and believed the information from, the generous trainer.

Active lying took a most interesting form, pointing—a behavior that chimpanzees never demonstrate in the wild. The chimpanzees were not taught to point; nevertheless, pointing emerged in all four animals. Perhaps all chimpanzees are able to point but do not manifest the behavior spontaneously except under extreme provocation. Pointing could have emerged in the laboratory simply because our experimental situation provided greater impetus for lying than is encountered in the wild.

Even though pointing emerged spontaneously in all four animals (something we never anticipated), the action never generalized (a further surprise). No chimpanzee ever pointed to another when in the compound, hall, or elsewhere. Pointing remained confined to the small experimental room in which it first emerged. Never did pointing develop into a general communicative act.

A child of about eleven months of age is already pointing. Often the child points when calling the parent's attention to a favorite object. Shouting the name "iff! iff!" for a fish swimming in a bowl, or "tuk! tuk!" for a truck moving down the street, the child looks imploringly into the parent's eyes (to verify the parent's gaze) and directs the mother's attention by pointing to and naming an object. Children are extremely keen about sharing their many excitements—far more keen than are chimpanzees—and they are well equipped to share their interests. Children point, look, speak, and they do so with remarkable coordination.

While the juvenile chimpanzees showed little evidence of being able to attribute intention to others, Sarah demonstrated some evidence for attributing intention in her series of video experiments. Shown videotapes of an actor facing a number of problems, Sarah indicated how well she understood the notion of *problem* by consistently choosing correct solutions. Since videotapes do not depict problems, but only a series of events, it was up to Sarah to interpret each series as a problem. For instance, pictures showing a person jumping up and down beneath a bunch of bananas have no particular meaning. Only when we analyze the pictures as depicting a hungry individual who is trying to reach some food, do we see a problem—and are then in a position to choose a particular solution. To find a problem in the video scenes requires the same kind of interpretation as is required to see action in the sequence: apple-blank-cut apple. Sarah was able to do both, while the non-language-trained animals were able to do neither.

When very young children (three-and-a-half years) are shown these videotapes, they do not pick solutions but choose, instead, pictures of objects that match some object seen in the videotape. Some children even select pictures of a TV set. Clearly, in their case, simply matching, and not the attribution of intention, is at work.

Sarah's success in attributing intention to others led us to ask what other states she might attribute. Could she attribute more knowledge to an older child than a younger one? To an adult? To a child rather than a chimpanzee? By the age of four, children can make distinctions of this kind, but not Sarah. Could Sarah distinguish between an individual who knew the answer to a problem and one who was merely guessing? Again, children, by the age of about six, can make this distinction, whereas Sarah does not.

Quite as the human body can act intentionally in more ways than can other bodies, so can the human mind attribute more states of mind. Perhaps intention is the most primitive of all mental states—the one state attributed to another if an individual can make any attributions at all. *Intention* is of course only *one* of myriad states that can be attributed. Children, we have seen, attribute not only intention, but also guessing, knowing, lying, hoping, wanting, and so on.

No child needs to attend school to be formally trained in the art of attribution. As with speech and the analysis of the world into causal relations, social attributions come naturally to the child. Children even graduate to the attribution of attribution, granting to others the capacity for making attributions. This is a step no chimpanzee can take, no matter how it is trained.

Mental Representation

If we took pictures of members of our family, then cut these pictures into parts, we could, if shown the nose, hair, chin, mouth, identify the owner of each facial portion. The smaller these pieces could be made and still ensure our accuracy, the more would this demonstrate that we had formed a good representation of our family members. Sarah evidently formed just such a clear representation of all the fruit included in her diet. For when shown stems, seeds, colors, outlines of shapes, and so on, she could identify the fruits of which they were a part. The other animals were also able to identify the parts, though not as reliably as Sarah.

If a chimpanzee could draw pictures, or outline maps, would its maps and pictures be different from those we produce? Chimpanzees form good mental representations of real fruit and actual space. Yet, they perform surprisingly poorly when not the *real* but our *representations* of fruit or space are shown to them. Chimpanzees have great difficulty seeing a correspondence between photos, pictures, TV images, maps, and dollhouses and the real objects and actual spaces they represent for us.

A two-year-old child who is shown (on television) where an item has been hidden in a room can find that item when it later enters the room. Or if shown the location of an item in a room, he or she can point to the item in a TV picture of the room. The two-year-old chimpanzee fails this problem (in both directions), so does the four-year-old, and so, probably, will the eight-year-old chimpanzee. Sarah, at about ten years of age, could do such problems, but with less accuracy than even the two-year-old child.

The difference in the performance cannot be explained by lack of experience, for Sarah has had considerable access to commercial television for over twenty years. Moreover, children can match pictures to objects (and vice versa) not only when very young, but even if reared without ever seeing pictures. Although chimpanzees can translate real-world objects and spaces into viable mental representations (using them in reasoning and problem solving) they are weak at doing what, in a sense, is the opposite—taking our versions of mental representations, in the form of photos, maps, dollhouses, TV pictures, and using them as a guide to the real world.

The chimpanzee's weakness is first suggested when it is given the task of matching pictures with the objects they represent. The animal is slow at learning to match the picture with its correct object, and vice versa. Initially, the animal makes a curious and consistent error; it matches objects to objects, and pictures to pictures. For instance, rather than match a picture of a banana with an actual banana, it matches a picture of a banana, with, perhaps, a picture of a shoe. Severely retarded children do a similar thing. This tendency suggests that the chimpanzee may be sensitive to (or dependent upon), the periphery or outlying boundary of an object—placing together pictures because their boundaries are alike. The mental representations the chimpanzee forms of objects and spaces may include a heavy reliance on "boundaries." It is quite possible that the TV images, dollhouses, and photos we have shown to the chimpanzees do not make the boundaries of their compound and rooms clear enough; so that, even though the animals form a good mental representation of the dollhouse, photo, and so on, they cannot link this representation to their representation of the actual compound and rooms. Whatever the source of the difficulty, it is clear that the chimpanzee has problems with the use of external representations—problems that children do not have.

So far, our examples of the representational system have been defined by the imaginal code, where representations have the quality of "pictures." That is, mental representations "look like" real objects and spaces. But mental representations can be of a totally different kind. In the abstract code, representations can take the form of "words," and words do not at all look like the things they represent. In fact, words are particularly interesting because they combine both imaginal and abstract codes.

When words are learned, they are initially associated with the imaginal code. For instance, if the word "cut" is taught with the use of an apple and a knife, the word becomes associated with an imaginal representation of this particular action. However, the word "cut" will also be associated with a more general condition—the separating of an object into parts—and this general condition is represented in the abstract code. Although we are taught words with specific examples, which are then represented by specific images, we also represent words abstractly. The child can use the word "cut" in situations that are completely different from that of cutting an apple with a knife, demonstrating that more than a specific image is associated with "cut." The child will transfer the abstract idea of cutting to include: the cutting of hair, paper dolls, and, eventually, classes in school—evidence for a dual system of representation in which the abstract code is particularly effective.

If we teach two children "cutting," one with apples and paring knives, the other with paper and scissors, the two would have no difficulty understanding one another's idea of "cutting." They could, in fact, understand a total stranger who mentioned just having had his hair cut. But if we teach two pigeons "color of," using apples with one and sky with the other, the birds will have no idea what the other means by "color." Having only an imaginal code, one bird will picture "color" in terms of red (apple), while the other will picture color in terms of blue (sky). Not having an abstract code, the birds are unable to represent color in any but the very specific way in which the word was first learned. The chimpanzee, however, has an abstract code. Sarah, who was taught "color of" with apple (red) and banana (yellow), transferred the abstract idea of color to grape (green) and chocolate (brown). She, like the young child, would be perfectly able to interpret the confusions over color experienced by the two pigeons.

Because it has an abstract code, the chimpanzee can be taught an artificial representational system—a language. However, the chimpanzee's language must be carefully distinguished from human language. The difference hinges on the comparison between the construction and the sentence. In a construction, we find a simple, one-to-one correspondence between words and the real-world items they refer to. And a careful, meticulous training procedure must be used to establish a relation between the order of words and the situation described by the words. In learning a language, children spontaneously observe a word order that they adhere to. Their sentences are not strictly a correspondence between words and real-world items, but are open to indefinitely many variations. We can clarify our notion of sentence by offering an analogy between conceptions such as those of "cutting" and the conceptions associated with sentences.

How do we know that someone has the abstract idea of cutting? By the fact that he can point to examples such as "cutting" an apple, clipping coupons, sawing a board, cutting hair, and so on. How do we know that someone has the abstract idea of the simple sentence: "The red rose is on the table?" By seeing the individual go out into the world and point to examples described by the sentence? Certainly not. One need never leave the room at all. We know someone has the abstract idea of a sentence when she can produce indefinitely many variations of the original statement, that is: "On the table is the red rose," "The rose on the table is red," "The table has on it a red rose," "A rose which is red is on the table," and so forth. The abstract idea of cutting is to its many real world examples quite the same as is the abstract idea of a sentence to its many possible forms.

The same language system that produced an upgrading of the ape's mind, has been, over the years, used with a wide variety of pathological human populations. Sarah's plastic language system has been taught to retarded and autistic children, who do not acquire language in the first place, as well as to adult stroke victims, who have lost all use of language. The training does not instill human language in the children or restore it in the adults. As with Sarah, the training provides constructions only. And yet, there seems to be every reason to take an optimistic view about the value of teaching language to those who have lost the capacity through an accident, or those who never acquired it naturally. If we can upgrade the mind of a chimpanzee, why not the mind of a human?

Upgrading the chimpanzee's mind by language training simulates, in a sense, a chapter in the history of our own species. Our ancestors, like the present-day ape, had the potential for abstract representation. The potential underwent substantial upgrading, not through training, of course, and not in eighteen months. But the change was not only in mental representation; the body itself changed. It became increasingly able to register intention. The will of each individual could be expressed not only in movements of the limbs and trunk, but in every possible nuance of face and voice. The mind kept pace. Attributing not only intention but myriad other states to other minds around it. Reasoning, too, expanded. Analogy and metaphor grew from simple, spontaneous sorting. Similarities that the child merely perceived in the world became abstractions that the adult attributed to the world.

Earlier we said that we hoped to understand the mind of the human. In order to do so, it was essential, we thought, to compare it with other minds, to see if other minds were a simpler version of the human mind. Did language, we wondered, completely transform the mind of the human? Or did it simply add another organ to a basic structure that was otherwise like the mental structure of all other species? After twenty years of research, we propose the existence of at least three kinds of minds.

Each mind must solve essentially the same general problems; analyzing the physical and social worlds and representing both worlds mentally. One kind of mind, probably shared by most species of nonprimates, specializes in imagery. It has neither abstract code nor language. It cannot recognize representations of action, nor does it make social attributions. Another kind, which humans share, has imagery, an abstract code, and language. The chimpanzee represents the third kind of mind, one which has, in addition to imagery, a capacity for abstract representation. Teaching language to this kind of mind does not confer human language, but appears to upgrade its ability to solve abstract problems.

Adding a human larynx to the ape would not make of it a human, nor would subtracting language from the human make of it an ape. Over vast periods of time, genetic changes transformed a creature with a mere potential for abstract representation into the present human. It is not language alone that separates the human mind from that of the chimpanzee.

Annotated References ■

Introduction: The Hairy Golem

There are many books and monographs that describe the chimpanzee in three different kinds of habitats: the wild, the human home (essentially as an adopted child), and the laboratory. While many of these references are not recent, they remain our best sources of information on chimpanzees.

On the chimpanzee in the wild:

Goodall, J. van L. (1971) *In the Shadow of Man.* Boston: Houghton Mifflin.

Reynolds, V. and Reynolds, F. (1965) Chimpanzees of the Bubongo forest. In Devore, I. (Ed.) *Primate behavior: Field studies of monkeys and apes.* New York: Holt, Rinehart and Winston.

Plooij, F. X. (1978) Some basic traits of language in wild chimpanzees? In Lock, A. (Ed.) *Action, Gesture, and Symbol.* New York: Academic Press.

Nishida, T. (1968) The social group of wild chimpanzees in the Mahali Mountains. *Primates* 9:167–224.

On the chimpanzee as an adopted child:

Kellogg, W. N. and Kellogg, L. A. (1933) *The Ape and the Child: A Study of Environmental Influence Upon Early Behavior.* New York: McGraw-Hill.

Ladygina-Kohts, N. N. (1935) *Infant Ape and Human Child.* Moscow: Museum Darwinianum (though written in Russian and never translated, the book contains a useful English summary).

Hayes, Catherine (1951) *The Ape in Our House.* New York: Harper.

Hayes, K. L. and Nissen, C. H. (1971) Higher mental functions of a home-raised chimpanzee. In Schrier, A. M. and Stollnitz, F. (Eds.) *Behavior of Nonhuman Primates.* New York: Academic Press, pp. 60–114.

Descriptions of chimpanzees, studied in the laboratory:

Köhler, W. (1925) *The Mentality of Apes.* London: Kegan, P.

Jacobsen, C. F., Jacobsen, M. M. and Yoshioka, J. G. (1932) Development of an infant chimpanzee during her first year. *Comparative Psychological Monographs* 9(1), 1–94.

Yerkes, R. M. *Chimpanzees.* (1943) New Haven: Yale University Press.

Premack, A. J. (1975) *Why Chimps Can Read.* New York: Harper.

Premack, D. (1976) *Intelligence in Ape and Man.* Hillsdale, N.J.: Erlbaum.

On the evolution of humans and apes:

King, M. C. and Wilson, A. C. (1975) Evolution at two levels in humans and chimpanzees. *Science,* 188:107–16.

There are presently many books on language and the psychology of language (as there were not in 1954). Of them, the two having very different perspectives and likely to remain of interest for many years are:

Skinner, B. F. (1957) *Verbal Behavior.* New York: Appleton-Century-Crofts.

Chomsky, N. (1957) *Syntactic Structures.* The Hague: Mouton & Co.

Other relevent books on language include: Searle, J. R. (1969) *Speech Acts.* London: Cambridge University Press; Lenneberg, E. H. (1967) *Biological foundations of language.* New York: Wiley & Sons. Of the general texts on the psychology of language three notable ones are:

Fodor, J. A., Bever, T. G., and Garrett, M. F. (1974) *The psychology of language.* New York: McGraw-Hill.

Miller, G. (1981) *Language and Speech.* W. H. Freeman and Company.

DeVilliers, J. G. and DeVilliers, P. A. (1978) *Language acquisition.* Cambridge: Mass.: Harvard University Press.

Chapter 1: A Language Designed for an Ape

In order to form an association between two specific items, must an individual first classify the item? This assumption was suggested to us by data from the chimpanzees: the length of time the animals required to learn their first "words". Even after 500 trials, the chimpanzee may not know the actual word for each fruit. However, it does know the following at the language board: *use* the pieces of plastic; *take* the slices of fruit; and *avoid* both using or taking unfamiliar words and fruit. For example, the chimpanzee appears to learn "any piece of plastic can be used to get any fruit"; that is, it first forms an association between the class of fruit and the class of words. Only after this association is formed does the animal go on to learn "blue triangle goes with apple". Is it possible to associate items that have not been classified? Is there a predilection or a necessity to learn in a hierarchical manner in the primate? These issues are discussed in: Premack, D. (1973) Cognitive principles? In McGuigan, F. J. and Lumsden, D. B. (Eds.) *Contemporary approaches to conditioning and learning*. Washington, D.C.: Winston Press.

Chapter 2: From Simple Judgments to Analogies

Young children touch alike items in the same interval of time (temporal sorting), put them in the same place (spatial sorting), and find one-to-one correspondences between two groups of objects. For example, given a set of marbles and cups, the child places a marble in each cup; given a group of dolls and chairs, sets a doll in each chair. Susan Sugarman discusses spontaneous organization in young children in: *Children's early thought: developments in classification*. N. Y.: Cambridge University Press, in press.

Although young chimpanzees can perform both temporal and spatial sorting, they do not then place groups of items into one-to-one correspondence. The ape can be trained to the task (as Sarah was, see *Intelligence in Ape and Man*) and can transfer the task to new items, but the ape cannot manage the task spontaneously, as the child can. Since language consists of a mapping of one system (words) onto another (units of meaning)—a version of the correspondence task—we are tempted to associate both the presence of natural language in the child with its tendency to form correspondences, and the absence of language in the ape with its *failure* to form such correspondences. Although an intriguing hypothesis, it remains without proof.

The fascinating evidence for the presence of natural concepts in the pigeon was first reported in: R. J. Herrnstein and D. H. Loveland. "Complex visual concept in the pigeon." *Science*, 1964, 146, 549–51.

Sarah's data on analogies come from: Gillan, D. J., Premack, D., and Woodruff, G. Reasoning in the chimpanzee: I. Analogical reasoning. *Journal of Experimental Psychology: Animal Behavior Processes*, 1981, 7, 1–17. In these tests, Sarah either completes an incomplete analogy or judges a pair of relations to be "same" or "different". In more recent work, David Oden found Sarah capable of building analogies from "scratch". When given an analogy board, along with four cards, each containing a figure of the kind described in the perceptual analogies, Sarah forms analogies correctly about 62 percent of the time. (The cards can be arranged on the board in 24 different ways, only 8 of them analogies, making chance 33%.)

Chapter 3: Does the Ape Believe You Have Intentions?

An especially readable account of what philosophers of mind mean by intention, belief, and similar topics can be found in: D. C. Dennett. "Intentional Systems." *The Journal of Philosophy*, 1971, 68 (no. 4), 87–106.

Nice examples of the development of intentional behavior in the chimpanzee can be found in the reference from Plooij, given above. An important comparative account of limitations in the transition from elicited to voluntary behavior in monkey, ape, and child can be found in: Chevalier-Skilnikoff, S. (1976) The ontogeny of primate intelligence and its implications for communication potential: A preliminary report, in Harnad, S. R., Steklis, H. D., and Lancaster, J. (Eds.) *Origins and evolution of language and speech*. New York: New York Academy of Sciences. John Flavell discusses development of social attributions in the child in: "The development of inferences about others," in Mischel, T. (Ed.) (1974) *Understanding Other Persons*. Oxford, England: Blackwell, Basel and Mott.

The data on "lying" in the juvenile chimpanzees are taken from: Woodruff, G. and Premack, D. "Intentional communication in the chimpanzee: The development of deception. *Cognition*, 1979, 7, 333–62.

The research on the attribution of intention, using Sarah as the subject, can be found in: Premack, D. and Woodruff, G. Does the chimpanzee have a theory of mind? *The Behavioral and Brain Sciences*, 1978, 4, 515–26.

Chapter 4: The Ape and Its Physical World

The data on the children's performance on conservation of number, liquid, and solid quantity (using Sarah's plastic words for same and different) was collected by Julie Rosenberger for an undergraduate research project at the University of Pennsylvania in 1978.

Chapter 6: Translating "Pictures" into "Words" and Vice Versa

The ability of a child to identify objects from their pictures, even though never having previously been shown pictures, is documented in: Hochberg, Julian and Brooks, V. *Pictorial Recognition As an Unlearned Ability: A Study of One Child's Performance. American Journal of Psychology,* 1962, vol. 75, pp. 624–28. The child was, of course, Hochberg's own daughter.

Guy Woodruff is mainly responsible for attempting to teach map reading to the juvenile chimpanzees. Guy spent four fruitful years at the laboratory collecting more data than we will likely be able to publish. R. Kellogg, in her book *What Children Scribble and Why* (published by the author, San Francisco, 1955), has provided the instructive cross-cultural data on the development of representational drawing in children.

There are many books on reasoning in the child, all of them influenced by Piaget's thinking. A classic is: Inhelder, B. and Piaget, J. *The early growth of logic in the child: Classification and seriation.* New York: Harper and Row. Dan Osherson's *Logical Abilities in Children,* vol. I (Hillsdale, N.J.: Erlbaum, 1974) offers, in addition, a criticism of Piaget.

Chapter 7: Who Has Language?

The data on subject deletion in the young child come from sentence protocols collected on both normal and retarded children by Anne Fowler; we are indebted to her for allowing us to refer to them here. How do children acquire language? In attempting to answer this extremely complex question, some place the emphasis on the mother (environment), others on genetic (maturational) factors. Naturally, on a matter of this complexity, there is a wide disparity of opinion, and although the recommendations that follow reflect this diversity, they have in common thoughtfulness and clarity:

Bloom, L. *One word at a time,* The Hague: Mouton, 1973.
Bowerman, M. Reorganizational processes in lexical and syntactic development. In E. Wanner and L. R. Gleitman (Eds.) *Language acquisition: The state of the art.* New York: Cambridge University Press, 1982.
Braine, M. D. S. Children's first word combinations, *Monographs of the Society for Research in Child Development,* 1976, 41, Serial no. 164.
Newport, E. L., Gleitman, H., and Gleitman, L. R. Mother, I'd rather do

it myself: Some effects and non-effects of maternal speech style. In C. E. Snow and C. A. Ferguson (Eds.) *Talking to children: Language input and acquisition*. New York: Cambridge University Press, 1977.

Pinker, S. Formal models of language learning. *Cognition*, 1979, 7, 217–83.

Slobin, D. I. Universal and particular in the acquisition of language. In Wanner, E., and Gleitman, L. R. (Eds.) *Language acquisition: The state of the art*.

Chapter 8: Characteristics of an Upgraded Mind

The juvenile chimpanzee's initial match-to-sample performance when first tested on plants in the compound come from data collected by Guy Woodruff.

The data on transitive inference comes from: Gillan, Doug. Reasoning in the chimpanzee: Transitive inference. *Journal of Experimental Psychology: Animal Behavior Processes*, 1981, 7, 105–64. B. McGonigle and M. Chalmers have reported similar behavior in monkeys ("Are monkeys logical?" *Nature*, 1977, 267, 694–96).

The effects of pedagogy on cognition come from: Cole, M., Gay, J., Glick, J. A., and Sharp, D. W. (1971) *The cultural context of learning and thinking*. New York: Basic Books. The effects of literacy on cognition come from Scribner, Sylvia and Cole, Michael. (1981) *The Psychology of Literacy*. Harvard University Press.

Conclusion

A wide-ranging discussion of the approaches used in teaching language to languageless humans—those thousands who have failed to acquire language and those who have lost their use of language as a result of neurological trauma—can be found in: Schiefelbusch, R. L. and Lloyd, L. L. (Eds.) (1974) *Language perspectives-acquisition, retardation, and intervention*. Baltimore, Md.: University Park Press.

Sarah was more consistent in judging amount—of both liquid and solid—than in judging number. She matched "small" numbers correctly, as do most chimpanzees, but sometimes failed to make same/different judgments about them. For the child, number is a dominant feature, as discussed in the notable work: Gelman, Rochel and Gallistel, Randy (1978) *The child's understanding of number*. Cambridge: Harvard University Press.

Talking (at least, thinking) chimpanzees have made their debut in recent fiction. Vercors, in *You Shall Know Them*, wonders whether religion (or even a religious trinket) is not at the heart of being human; and whether a chimpanzee, attached to such a trinket, is not a "person." The *Poison Oracle*, by Peter Dickinson, features a chimpanzee who talks with plastic words and happens to be the sole witness to a murder. In *God's Grace*, Bernard Malamud makes a classic error: his chimpanzee is essentially a human-without-language, for when provided a larynx and a little training, the animal abruptly acquires a human mind. As we have seen, the chimpanzee's is a different kind of mind.

Index

chimpanzees *(continued)*
  abstract art attributed to,
      108
  analogies made by, 43–47
  "and/or" in, 109–13
  anthropomorphized, 5
  biting by, 11
  body postures of, 54–55
  cheating by, *see* social cues
  constructions formed by,
      118–19, 123
  domesticated animals vs., 11
  drawing by, 108
  facial expressions in, 54
  functions of objects learned
      by, 22, 105–6, 108
  gestures of, 52, 54–55
  memory of, 120–22
  mutual staring of, 2, 10
  natural diet of, 6
  natural potential of, 139–40
  numbers and, 78, 97, 143
  painting by, 108
  as pets, 4–5
  photo-object matching of,
      101, 147–48
  pointing by, 52, 55
  as questioners, 24, 29
  restlessness of, 12
  social exchanges with,
      17–18
  sorting by, 35–36
  speed of, 6
  syntactic distinctions by, 115
  vocalization in, 54
  weaning age of, 7
chimpanzees vs. humans:
  anatomy in, 2
  in attributing attribution of
      intentions, 66–67
  chromosomes in, 2
  spontaneous categorization
      in, 35–36

  in tests of conservation, 78,
      81–82, 128, 143
chimpanzees vs. monkeys:
  in action tests, 142
  gaze reactions of, 2
  observational learning in, 70
chimpanzees vs. pigeons:
  abstract systems and, 100,
      136
  in match-to-sample tests,
      38–39, 136
chimpanzees vs. rats:
  abstract systems and, 100,
      136
  in match-to-sample tests,
      136
chimpanzee-trainer relation-
      ship:
  grooming in, 6
  maternal ties in, 7–9
  social cues in, 83–90
  social exchanges in, 17–18
Clever Hans procedure, 84,
      85
"color of," 87, 138
conservation, tests of, 77–82
  children vs. chimpanzees in,
      78, 81–82, 128, 143
  with liquids, 78–81, 128,
      143
  with numbers, 78, 143
  with solids, 78, 81, 128,
      143
constructions:
  new words and completion
      of, 138
  sentences vs., 118
"cut," 20, 148–49

deception, *see* lying
"different," 25, 40, 92–93
double-blind procedure, 84,
      95–96, 97

DATE DUE

MAY 95

DEC

OCT

NOV

GA

PRINTED IN U.S.A.